Sound S

to achieving a more musical Hi-Fi system

Russ Andrews

Solutions for better music & movies™

A Russ Andrews Publication

Editors
John Armer
Kirsteen Andrews

Proofing Team
Gary Hartley
Dalia Remeikyte

Design
Sarah Garstang

Illustration
Sarah Garstang
Jo Keegan

First published in Great Britain in 2006
by Russ Andrews Accessories Ltd

Printed by Badger Press
Bowness-on-Windermere, UK.
+44 1539 445399

ISBN
0-9554275-0-9
978-0-9554275-0-3

List of Diagrams

Table of Contents

Foreword

Honest to goodness: the first person to 'get it', when I decided to build my dream listening room was Russ Andrews. He knew immediately what I needed to make the room 'politically correct': wiring, AC distribution layout, socket types. And I didn't really need to phone him: it was all in his catalogue.

Russ was the first person I turned to because I'd known him for decades, and knew his tenacity and focus. He provided the cables for one of the circuits, some mains-capable Kimber that turned out to be the most 'neutral' of the various circuits. But I shouldn't have been surprised, because Russ was one of the first tweakers in the UK to go that extra mile, to try to find out *why* stuff does what it does.

He never, for example, took tables and supports for granted, never accepted from the outset that sheer mass was enough. Instead, Russ developed Torlyte tables, which behaved quite differently from the run-of-the-mill heavy metal, and which enabled users to fine-tune their systems.

Eschewing the truly silly solutions that have come and gone, Russ Andrews – via Russ Andrews Accessories - has refined the procedures, discovered the shortcuts. *Sound Solutions* is full of advice and free tips, all of which came about through listening and experimentation rather than casting rune stones. Russ wrote this book because he's one of us: a music-loving hi-fi enthusiast eager to get the most music out of a system.

Ken Kessler

Preface

This book sets out to describe my unique approach to upgrading your Hi-Fi system. It shares with you the knowledge I have gained over many years in the Hi-Fi industry as a retailer, research and development consultant, designer and enthusiast.

The key to my approach is that the infrastructure supporting your Hi-Fi system – the cables, supports and even the room it is in – is **every bit as important** as the Hi-Fi kit itself. *Sound Solutions* passes on the secrets and techniques (and the common upgrading pitfalls so you can avoid them) I have discovered over the years that make it possible to get truly enjoyable music from any system, be it large or small, 'High End' or humble. And it's *music* that is the driving force behind writing this book. Music is the real reason to own a Hi-Fi and it's a sad fact that without applying the solutions I offer here, many systems rob the life – the *musicality* – from music.

Throughout the book you'll see references to cables from the US manufacturer Kimber Kable. I mention them not as a shameless 'plug' (my company Russ Andrews Accessories is the UK importer and distributor for Kimber Kable) but simply because I'm so familiar with them and they illustrate some of the points I make very well. Many of the concepts that I put forward and tips that I suggest can be applied using other manufacturer's products as well as those from Kimber.

I am indebted to Ivor Tiefenbrun for enlightening me to the concept of 'Rhythm and Timing' when he first demonstrated the LP12 turntable to me in 1973. I knew some things sounded more musical than others but didn't know why. He identified for me what it was about live music that really differentiated it from reproduced. I can only repay him by passing it on to you. I am further indebted to my editors John Armer and my daughter Kirsteen, and our Graphic Designer Sarah Garstang for making the book readable, coherent and published on time, and also to Ken Kessler for his generous Foreword.

Chapter One
The Music and The Magic

Unconventional Wisdom

Don't imagine you have to spend hundreds (or even thousands) to be able to achieve truly enjoyable sound from your system. Just spending a small amount getting the fundamentals right will bring huge rewards in increased performance and enjoyment. Follow my 'Upgrade Path' (page 19) and you will sort out the most important, fundamental things for your system, you will do it in a way that will give you the biggest improvement for your money at each stage, and you will end up with a sound that makes you sigh with satisfaction!

Let me say, right from the outset, that what you read here is not the conventionally accepted wisdom of how to put together and upgrade a successful Hi-Fi system. If the conventional recipe worked, you probably wouldn't be reading this and I wouldn't have needed to write it because everyone would have a great Hi-Fi system that gave them endless musical pleasure.

The sad truth is that most people end up with a collection of well-reviewed hardware that gives them endless displeasure. Anyone with an ounce of common sense would give up and buy a midi system! But no, we're determined not to be beaten so we persist with the charade, buying ever more expensive components because the magazines and Hi-Fi dealers tell us that's the answer.

It's a bit like owning a second hand car. You want to sell it, but it needs repairing first, after which it owes you more than it's worth and you can't afford to sell it until you've had some use out of the repair. At which point it goes faulty again so you have to invest more money. An endless trap; just like the Hi-Fi trap. People often say to me *"I've got so much money invested in a system that's worth very little on the second hand market that I can't afford to sell it and start again. Can you help me get music out of it because no one else can?"*

The answer is *"Yes, I can",* but it requires a completely different approach to the problem and, inevitably, more money must be invested in the system.

My quest for the answer to the question of why a collection of highly regarded hardware doesn't always add up to a musical system started in 1972. I then achieved my ambition of putting together what was regarded as the best system of its day. I had a Thorens TD124 turntable with SME arm and Shure V15 cartridge, a Quad 33/303 pre and power amp and a pair of Spendor BC1 speakers. The result was deeply disappointing. It made a sound that I later described as 'Hi-Fi', without any musical merit whatever. I felt cheated and betrayed by the industry; I was a Hi-Fi dealer after all and I was selling this stuff to others! I was so upset I suppose, because it was so much worse than my old, home built, valve based system. A system I had sold, along with my sports car, to raise enough money to get married. (No regrets though, it was a terrific investment, Sue and I are still happily married!)

At first, of course, I questioned my own judgement and listening ability, but then it dawned on me that perhaps the industry had taken a wrong turning and lost its way. From that moment on I have been working to find out what went wrong and how to put it right.

I questioned everything (making me a very controversial figure in the process) and found that few people seemed to have any answers that worked. (Ivor Tiefenbrun's Linn LP12 turntable was one and Julian Vereker's Naim power amps were another, as many of you may remember). So I set to and researched everything for myself. Very early on I discovered how influential the mains quality was, starting a research programme that continues to this day.

Over the years I have become involved in the design of almost every part of a Hi-Fi system including, crucially, cables. It was this background in cable design and manufacture that enabled me to recognise the very special innovations that Ray Kimber of Kimber Kable had developed in both construction and manufacturing techniques when I first met him in 1985. My subsequent listening tests confirmed the huge leap forward that Ray had made. His cables had an honest, consistent, coherent design philosophy and I knew he would make them sound even better as time went on.

My development work included such arcane subjects as equipment supports and room acoustics and my work on the hardware showed that the sonic qualities of the individual electronic components had more influence on the sound than the circuit design itself.

All of this work has been innovative and ahead of accepted practice. Everything has been seen initially as revolutionary and controversial, but has then gradually been accepted. Among other things, I was the first to discuss and design solutions to improve mains quality; the first to discover cable directionality; the first to identify the importance of component quality; and the first to discover the importance of light, rigid stands for Hi-Fi equipment. These have not been isolated discoveries, but developmental steps on a long road to a full solution to the problem. Each step has been essential to the overall understanding of what is happening in a Hi-Fi system and how to get the best out of it.

The answer is not 'mechanistic': valves versus transistors or electrostatic versus dynamic speakers. Nor is it 'synergistic': the right 'combination' of hardware. The answer is that *everything* matters. Everything affects the sound, but some things are more important than others. At the beginning of the upgrade path the least important thing is the hardware. Only after the fundamentals of how and where you connect up and use the hardware have been dealt with will 'better' hardware be worth spending money on. Until then you are only hearing a fraction of the potential of what you have already invested in: the hardware you already own.

I have developed an approach to upgrading that I have called 'The Upgrade Path' (see page 19). It describes five fundamental upgrade steps - starting with the mains supply and working through to the room acoustics - that will enable you to realise the true potential of your system. Following this path ensures you sort out the most important things in the right order to avoid unwittingly hiding problems by treating symptoms (more on this later), and that you will benefit from the biggest improvement for your money at each stage.

Get those fundamentals right, and a system of inexpensive, modest hardware will give more pleasure and enjoyment than the very best hardware just connected up in the usual way.

Most importantly, following the 'Upgrade Path' technique I recommend ensures that any system, whether it is a portable stereo or a £50,000 system, will deliver its full musical potential. Don't imagine you have to spend hundreds on your system to be able to follow my advice and achieve truly enjoyable sound from it. Just spending a moderate amount on mains and speaker cables for your mini system will bring huge rewards in increased performance and enjoyment.

A customer reported an amusing story. He accidentally damaged both his amplifier and his speakers, so whilst they were away for repair he dug out his old music centre. He connected it up with our least expensive, budget cables – our entry level mains cable and Kimber's 4PR speaker cable. He was shocked to discover that it was then more enjoyable to listen to than his Hi-Fi system had been before he blew it up.

It didn't have all the attributes of a Hi-Fi system of course, it was lacking in many ways but, crucially, he enjoyed the music more. This is the real objective. When it really comes down to it, I don't care how it's done or what does it, I'm in it for the emotional buzz that music gives, not pride of ownership of fashionable hardware.

We are right back where I started with my 1972 system. I now know how to get real musical enjoyment out of it – but it's taken since then to learn how to do it! Save yourself all that frustration, time and expense; read this book to share the knowledge gained from my research, follow the five steps in my 'Upgrade Path', and you'll soon be enjoying your system a whole lot more! You've got nothing to lose and everything to gain.

Listening to your System

So how do you form a mental picture of what you are trying to achieve? That's easy; you just listen to the natural world around you. Ask yourself what, for example, does a door closing sound like. Why does a real one sound so different from a recorded one? What aspects or elements of the sound are missing from the recording? Go out and listen to an audience clapping. Analyse the differences and the similarities between live and recorded clapping. No two people make the same clapping sound but there are identical elements; elements that clearly distinguish the live sound from the recorded one. Listen to the energy, dynamic range, impact, loudness, frequency range, cleanness, tonal character and also the room (or outdoor) response to the sound: echo, reverberation and decay.

Hi-Fi systems should be able to reproduce all these aspects of the sound. Sadly, they usually miss out most of it leaving a flat, lifeless travesty of the original. The object of a Hi-Fi system is to reproduce music (sounds) in a realistic, lifelike way, hence the name High-Fidelity! It must at least half convince you that the musicians are right there in front of you, playing in the room. Don't you wish your system could do that? Well, it can if you get the fundamentals, the context, right first.

The Magic of Musicality

There is an important aspect of real live music that is usually destroyed by a Hi-Fi system. For me it's the most important, most fundamental thing of all. It's called musicality.

Musicality is the rhythm and timing that puts emotional meaning into the music. Musicians either have it or don't have it. Maybe you hadn't realised why you listen to some musicians and not others. The secret of being a good musician is not just in playing the right notes in the right order but in the rhythm and timing. In fact a musician with great rhythm and timing can make mistakes and get away with it. Frank Sinatra, Louis Armstrong and Julian Bream come instantly to mind, but there are many others.

Hi-Fi systems can easily destroy the rhythm and timing in the music. I've heard both hardware and cables take the musicality out of the music in this way.

Take time to learn to listen for musicality because it's the easiest way to judge equipment or systems quickly and accurately. It isn't just a matter of whether the music makes you want to tap your foot, it's something you feel inside.

At Hi-Fi Shows I have used a portable stereo to demonstrate the benefits of our entry level mains cable over the original. Almost any CD is good for demonstrating the improvements in clarity, information, bass, etc., but I have a jazz recording that begins with a solo clarinet playing the intro to 'Mood Indigo'.

This short intro shows the improvement in musicality so obviously that people often laugh in amazement at the change. With the original mains cable the clarinetist sounds bored, rushing the notes and just glad to be at the end, but with a Russ Andrews mains cable he is clearly putting great skill into playing expressively and changing the pace to give meaning and emotion to it.

Find something like *'Mood Indigo'* above (the guitar intro *'Signe'* on *'Eric Clapton & Friends Unplugged'* is another good one) to use until you are really familiar with the effect, then you will find that you can hear it on any piece of music. The music is what it is all about. If the rhythm and timing are lost it doesn't matter how good the sound is, because you won't bother listening to it after the novelty has worn off. The importance of this distinction between a 'good' sound and a 'musical' sound is well illustrated by the comments from a fellow enthusiast, Mr Cosker after he'd had one of our mains extensions in his system for a couple of months: *"I'm amazed. It just keeps getting better, and now I would never part with it. It's the first time I can say I actually like my Naim system, rather than just being impressed by it. It's taken away that feeling of listening through slightly gritted teeth!"*

Musicality is the easiest way to judge any item of hardware, cables, mains, upgrades, accessories or systems. The best products have musicality designed in.

I can't over-emphasise the concept of musicality. It's fundamental to the reason for owning a Hi-Fi system - to play music. What musicality does for your enjoyment and listening pleasure is, quite simply, magical!

Learning to Listen

It's all very easy for me to say 'listen to the music' but, sadly, many enthusiasts have forgotten how to do it. I say 'forgotten' because non-Hi-Fi enthusiasts do it naturally. They don't, of course, use Hi-Fi terms to describe what they hear but they have no difficulty describing it in their own terms.

In fact they can say some very interesting and revealing things about good sounds. I once played a live recording of a rock concert to a young guy who was doing some creative packaging work for us. He sat there stunned with his mouth open. *"I was at that concert,"* he said, *"that's exactly what it was like being there. I didn't know you could do that!"* This kind of response to my system makes me want to jump up and down shouting *"Yes, Yes, YES!".*

Unfortunately, many Hi-Fi enthusiasts hearing the same system seem to miss the point entirely. They say things like, and I quote, *"Yes, but I think you've got the tonal character of these cellos wrong and I can't hear enough treble. Aren't your speakers much too far apart?"* Needless to say, this particular 'Hi-Fi Buff' had many thousands of pounds tied up in his Hi-Fi system and was always unhappy with the sound. (I suggested he sold his system and put the money into concert tickets, but he wasn't amused!)

If you can see some of yourself in the 'Hi-Fi Buff' don't panic, you're not alone... listen to what I am saying, put it into practice and you can put your system right and enjoy the music.

Way back in the early 1970s when I was a new eager young Hi-Fi dealer I demonstrated a pair of speakers to a couple who wanted to upgrade (well, he wanted to upgrade!).

I played them the hot new speaker of the day, the B&W DM3. She listened for a short while and then said, "What's that nasty fizzing noise?" In my naive enthusiasm I explained that that was high frequency extension, blah blah. She cut me off in mid-Hi-Fi babble and said, "I don't care what it is, it's a row and I'm not going to listen to it".

After the couple had left, I sat quietly in the dem room and thought about the failed dem. Three things became clear.

First, she was absolutely right about the treble - it was a row. And second, people who know nothing about Hi-Fi have a distinct advantage because their instinctive good judgment isn't undermined by the reputation of the products, reviews, specifications, price, or techno 'babble'.

Third, and most importantly, I knew that I had to work out a way I could do that for myself.

It took a while, but I eventually worked out how to listen quickly and accurately. I realised that my preoccupation with the reputation, reviews, specs, etc. of the equipment was getting in the way. These things were completely irrelevant because the only thing that mattered was what it sounded like.

I analysed my listening technique. I was using the usual defensive Hi-Fi tricks of concentrating hard on one small part of the sound to hear how it changed on an A/B comparison. Record surface noise was (and still is) a popular one, or an instrument or voice.

It's a bit like judging a painting by looking only at individual brush strokes with a magnifying glass - it tells you nothing useful about the painting itself. You have to stand right back from it to do that. And so it is with Hi-Fi. If you can trick your mind into mentally standing back and listening to the overall sound - the big picture - you hear everything quickly and easily.

When you are doing A/B comparisons for yourself try not to put yourself in the 'hot seat'. Create situations where you are a 'bystander'.

Everything that is good or bad, right or wrong about a sound is often more easily heard from outside the listening room by someone who is not concentrating on the sound. They hear the big picture because it's so obvious when you can stand back from it. The trick is in not concentrating, not trying too hard.

A customer once phoned me after a weekend listening session of some new speaker cable. He said he spent most of the Saturday, whilst his wife was out shopping, swapping from the old to the new cable and back again. He thoroughly confused himself, he no longer knew which was which let alone whether one was better.

His wife arrived home so he stopped switching back and forth (it always irritated her) and put the new cable in one more time. She stuck her head round the door and said, "That's much better, what have you just done?" "Oh," he said, "I've just put in that new cable I bought." "Brilliant," she said, "well done!"

He told me he was gob-smacked. How come his wife, who knows nothing about Hi-Fi and was in the kitchen putting the shopping away, could hear so clearly what he couldn't hear in the Hi-Fi room? The reason was that as a bystander she was able to hear the big picture.

Educating your ear in music or Hi-Fi quality is just like educating your palate in food; it takes time, patience and a willingness to experiment and learn.

Obviously a person brought up on junk food has a very limited experience and taste in food. They have never had the opportunity to develop a 'palate' because they have never been exposed to what we call 'real' food. When presented with new, interesting food they may only accept it covered in ketchup! But taking time to educate your ears (or palate for that matter!) will pay dividends in the long run.

The Importance of the Infrastructure

Before we go any further, it is important that you understand the impact of what I call the 'infrastructure' on the performance of your system. A Hi-Fi system is often described as a chain, starting at the music source and finishing at the loudspeakers, and it is said that the sound can only be as good as the weakest link. This can be a misleading analogy. Of course, a system has a start and a finish, but it does not start with the source and end with the speakers – it's far more complex than that! The chain actually starts with the mains and ends with the room environment.

The start is the mains because that's where the energy comes from that makes the system work. The quality of that energy is fundamental to the quality of sound the system produces. If the mains quality is poor then the sound quality will suffer. The end of the system is the room because the performance (sound) of the room has a profound effect on the performance of the system. There are several places where the system is open to feedback: the loudspeaker, the integrated amp (or the pre and power amps if you have them), the source (e.g. CD player) and the mains.

Looking at the **Hi-Fi Chain** diagram (see Diagram 1.1 opposite) it quickly becomes clear that a Hi-Fi system is indeed a very complex 'system' with many 'feedback inputs'. As you can see from the diagram and explanation opposite, there is acoustic feedback from the room getting into every part of the system. The feedback energy is now part of the signal and has distorted it in several ways. It has added colouration (resonance) to the signal and, most destructive of all, it has degraded the rhythm and timing of the music because the feedback energy is the same music but delayed in time. It is a time smear that robs the music of its musicality and naturalness. This explains why reducing these feedback inputs has such a massive knock-on effect on the performance of the system, and why each of my solutions has a disproportionately large effect. It also explains why changing one item can change the way the whole system works. This is why I advocate taking a 'holistic' view of the system infrastructure within its environment.

The Hi-Fi Chain

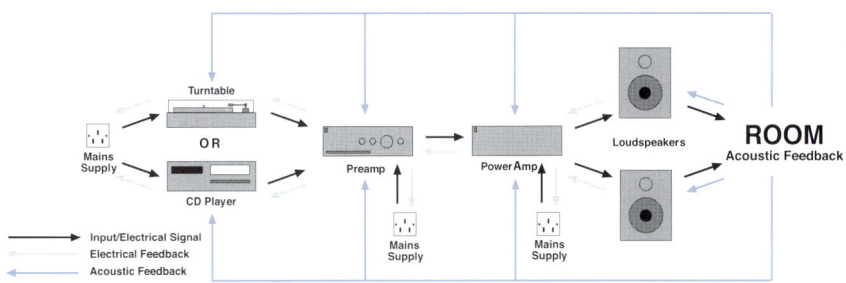

Diagram 1.1: Electrical and acoustic feedback in a Hi-Fi system.

1.The Loudspeaker

A loudspeaker is a two-way device. It works nearly as well as a microphone as it does a speaker. (Intercoms nearly always use the speaker as the microphone as well). In a Hi-Fi speaker the microphone effect is partially damped by the crossover and amplifier but it is still working as a microphone at the same time as it is acting as a loudspeaker! What is fed back into the amplifier is the difference between the input signal and the room acoustic feedback. This is a major reason why loudspeakers seem so very room and position dependent. The acoustic feedback becomes electrical feedback into the amplifier that colours and degrades the sound.

2.The Amplifier

The amplifier output stage 'sees' the feedback signal plus the RFI (Radio Frequency Interference - of which more later) on the speaker cable. The feedback loop does its job and changes the input to match the output, the protection circuits think the RFI is a fault condition and partially closes down the output stage in retaliation. The result is distortion that is most of the 'overload distortion' sounds you get when you turn the volume 'too high'. Get rid of the RFI and acoustic feedback problem and you will be able to turn the volume practically 'flat out' before real 'overload distortion' starts.

3.The Preamp

The feedback signal in the power amp doesn't just affect the output stage, it feeds right back to the input of the preamp. Remember, everything is connected together and works in both directions! The different signal on the preamp output (from the power amp) is fed back to its input (technically called *Back EMF*). Added to the back EMF is RFI picked up by the pre to power interconnects. Remember that coax cables are designed as good RF aerials and conductors so you can see

15

how unsuitable they are as audio connection cables. The Back EMF and RFI change and pollute the audio signal. (For more on EMF see page 44.)

4. The Source

As with the preamp, the Back EMF signal feeds through to the input, polluting the signal. Turntables are particularly prone to acoustic feedback. The cartridge turns any vibration into an electrical signal. Acoustic feedback from the room vibrates all the electronic components. These components are microphonic turning the vibrations into electrical signals that mix in with the audio signal.

5. The Mains

A portion of the signal passing through every piece of electronics in the system gets back into the mains and therefore into every other piece as a pollutant.

The 5 Keys to Successful Upgrading

Over the years I have discovered many common upgrading pitfalls that can lead off the 'path to Hi-Fi Nirvana' and set you firmly on the road to wasted money and ultimate disenchantment with your system. Here are my keys to avoiding those common pitfalls and so ensuring you achieve your goal!

1. Trust Your Own Ears!

When you change a component (hardware or cable) you get not one but two things to listen to. You get the sound of the new item and you get a new view of the sound of the system itself. You can only hear each item through all the others that are now interacting slightly differently.

Every item (and every component within every item) has a voice of its own within the overall sound. With enough comparative listening experience you can hear each individual voice the way a choirmaster can hear each voice in his or her choir. It sounds far-fetched and impossible but it's really no more difficult than recognising a person's voice. Anyone can do that, it is a skill we all have.

It's only impossible if you believe the old Hi-Fi propaganda that we can't trust our ears; that aural memory is very short and fools us all the time. Nothing is further from the truth. If you trust your ears, listen to them and believe what you hear, you will find that aural memory is very accurate and persists for many years.

I admit I have an advantage because not only have I over 35 years experience of listening to a huge range of Hi-Fi equipment, but I have also worked on the design and development of all parts of a Hi-Fi system. When I was working on cartridge development, for example, I was able to compare the sound of several

*different diamond tip shapes, cantilever materials, body plastics, etc. so
accurately that I could listen to any cartridge and make a good guess at not
only its make and model, but also at what parts it was made from.*

*Years ago a friend of mine, David McDowel, who then worked for Quad, played
me a tape recording he had made to prove once and for all that every turntable
sounded the same when used under controlled conditions. He had set up four
different turntables with the same arm and cartridge playing the same disc.
I was able not only to identify each turntable instantly (he hadn't said which four
he had used) but the arm and cartridge as well. I don't think he believed it - he
was sure that I had prior knowledge!*

2. Learn to Distinguish between a 'Symptom' and a 'Root Cause'

Every upgrade you make to your system - whether it be a cable, support or
hardware - will change your view of the music. The classic trap when setting up
a system and choosing cables is to unwittingly hide problems by treating
symptoms. A speaker cable that hides overload distortion can sound 'better' in an
A/B comparison than a more accurate and revealing cable. The problem is that
any better component not only improves the music but also inevitably reveals
problems not heard before. These are problems that were masked by the previous
inferior component.

So how do you tell whether an upgrade is better but has revealed other problems,
or whether the upgrade is a step in the wrong direction? Unfortunately there are
no short cuts, this is something that can only really come from years of experience.

This is the reason why I advise you to follow my 'Upgrade Path' (see page 19); do so
and you will cure the cause of each problem rather than treat symptoms. If you
don't start at the beginning and cure the cause of a problem you quickly end up
going in the wrong direction with your system.

That's why this upgrading process can seem so complicated and variable. And
that's why I sell and enthuse about Kimber Kable. Used with understanding, its
unique qualities solve problems and allow any system to deliver enjoyable music.

Remember:
- The root cause of a problem will often have several symptoms associated
  with it - not all of them evident at the same time.
- A symptom is not a problem in itself - it is merely a clue to a root cause of
  the problem.
- If you simply try and fix the symptom, the root cause is still there degrading

the sound; and the fix itself will cause other symptoms.
- Always use symptoms as clues to find the root cause - fix that and the symptom disappears.

Some examples of symptoms and causes:

Way back in the late seventies I sold a customer an LP12 turntable, Grace arm and Supex cartridge. He owned a big Pioneer receiver and wanted a Naim 32/250 combination. He asked me to demonstrate the Naim to him.

I connected it into his system and put on a record. He was shocked, it sounded terrible. What I realised, however, was that the Naim had revealed that the Supex cartridge was mistracking - a problem that had been inaudible through the Pioneer. A small adjustment to tracking pressure and the sound was just as he hoped it would be. If I hadn't been present, he would have rejected the Naim and the root cause would perhaps have never been discovered.

A few years ago I bought an old Landrover from a chap who assured me that I should ignore the recommended 30psi tyre pressures and keep them at 40psi as the ride was much better. Driving the beast home, I became convinced that the ride wasn't right at all, so I stopped and reduced the tyre pressure to 30psi.

The ride then became totally unstable, but it was obvious that the cause was worn out shock absorbers. The higher tyre pressures served only to mask the real cause of the bad ride. With new shock absorbers and correct tyre pressure the ride was safe and stable.

In the early days of selling Kimber Kable, I sold a customer a Kimber interconnect to fit between his CD player and amplifier. He loved the bass and midrange but complained that the treble was too hard and bright. I was puzzled because I knew the cable itself didn't have that characteristic.

Investigation showed that RFI on the mains was the cause of the hard sound and a PowerKord™ fitted to the CD player solved the problem and improved quality all round. In other cases, the PowerKord™ didn't completely solve the problem, leading me to discover the overload problem and so launch the attenuated CD interconnect (but that's another story!)."

3. Follow my 'Upgrade Path'

My 'Upgrade Path' will help guide you through the upgrading process a step at a time, ensuring that you solve problems at the root cause and get the biggest improvement for your money at each stage.

Step 1: **Mains**

Step 2: **Interconnects**

Step 3: **Speaker Cables**

Step 4: **Equipment Supports**

Step 5: **Room Acoustics**

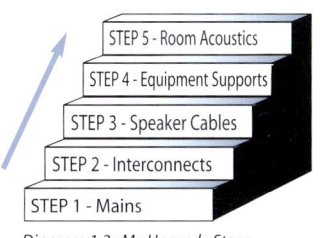

Diagram 1.2: My Upgrade Steps

In the real world you'll probably do a bit of each step at a time not the whole of Step 1 before the whole of Step 2, etc.

Please, whatever you do don't be put off starting because you aren't able to do the whole thing right now - in fact it can be much better to upgrade in stages, that way you get to really appreciate each improvement before you take the next step.

Experiencing my 'Upgrade Path'

When we sold Kimber Kable through a Dealer network, I went through my entire 'Upgrade Path' with two salesmen in a Hi-Fi shop. At every step along the 'Upgrade Path' they heard the improvements but were put off by the new problems revealed. They associated the problem with the upgrade instead of realising that it was just a hitherto masked symptom of the system itself.

Step 1: PowerBlock and mains cables. They loved the improved bass, soundstage, information, etc. but rejected the upgrade because the sound was brighter and a bit more aggressive. I explained that they could now hear the input overload more clearly. They were very sceptical of this explanation (their eyes said 'rubbish!!' but they were too polite to say it!).

Step 2: CD interconnect. It eliminated the overload and they were amazed, but they said that the dynamics had gone out of the music. I said that was false dynamics, a symptom of the overload. They said they didn't care what caused it - they liked it! I pointed to the speaker cable and blamed that for the lack of dynamics. More sceptical looks!

Step 3: 4TC/8TC Bi-wire speaker cable. Once again they were amazed. This time though, they didn't like the soundstage. I moved the speakers as far apart as the room would allow and gave them a sound stage the size of the dem room wall. They thought they had me with this one, because they were expecting a 'hole in the middle' they could push me through! They were wrong. They could not believe their eyes or their ears - I had them in the palm of my hand.

Step 4: Oak Cone Feet. I demonstrated the Oak Cones under the CD player and preamp and then went for the 'coup de grace' - ReVeel®/ReleeS®. I treated their favourite dem disc and they fell off their seats. One said, "where did that double bass come from - it wasn't there before!" I had them convinced at last.

I now use this upgrade demonstration in my Workshops – 'How to Upgrade your Hi-Fi System' and have produced a video of the demonstration. It works every time and when recorded on video you can hear all the changes clearly and obviously - even when played back through an ordinary mono TV speaker! That's because the differences are mostly musical improvements - rhythm and timing - not 'Hi-Fi' differences at all.

4. Don't Stop Until You Hit 'The Upgrade Plateau'!

Every day I face this problem of there being a downside to every upgrade. To help explain the process I describe the upgrade path as a staircase. The 'risers' are one or more upgrades that combine to reach a step or plateau of balanced enjoyable sound.

The 'risers' aren't always of equal height because it may take one upgrade or a combination of several working together to reach a tread or plateau. The plateaus are of unequal length depending on how big the previous upgrade was, the level of expectation and the amount of money available for spending on the Hi-Fi system.

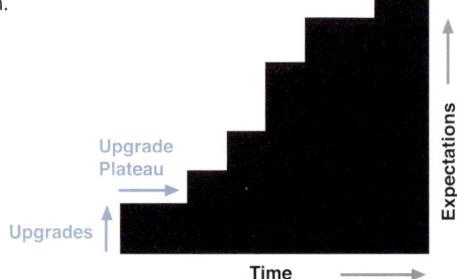

Diagram 1.3: The Upgrade Staircase.

To get the most out of the upgrade process, a clear understanding of the 'staircase effect' is essential. My advice is to go up the vertical upgrade step with enough upgrades combined to break through your current level of expectation. By that I mean reach a level of performance that makes the music sound better than you thought it could be, then stop upgrading, take time to enjoy the new sound of your music collection, buy new recordings and relax with the system. As time goes on the wonderful new performance will start to sound ordinary, you will hear problems you hadn't noticed before, etc. This is normal because your ears learn the new sound, stop being impressed by it and expect something better. It's now time for another upgrade step.

Upgrading your system's performance this way, a step at a time, gives you better value for money (because you have time to appreciate each improvement) and spreads the financial outlay over a longer period.

5. Rethink Your Budget

Many Hi-Fi magazines and Dealers will tell you that you should not spend more than 10% of the hardware budget on cables or you are wasting money. They believe in the Rule of Diminishing Returns! I have discovered a Rule of Increasing Returns that comes into play when you upgrade the right things in the right order. To achieve this you need to apply **my** 10% Rule. Spend 10% of your budget on the hardware and 90% getting the best out of it! (Okay, I may be exaggerating a little but not by as much as you might think.)

The standard '10% Rule' might sound fine, but it vastly underrates the fundamental importance of the cabling in a Hi-Fi system. There's no point spending money on your hardware and then never hearing its full potential because the cables you use add distortion and lose information. Think about it in terms of the size of improvement you'll gain rather than what you might 'expect' to pay. For example, if a £100 cable makes your £200 CD player sound better than a £500 CD player then that's excellent value for money by anyone's standards.

Achieving Your Goal

The most common cause of a 'failed' upgrade is in not doing enough in one stage to reach the next 'enjoyment plateau'. That's where failing to understand the difference between causes and symptoms comes in, especially when coupled with a 'distorted' expectation of what music should sound like. The salesmen in the Hi-Fi shop that I described earlier had both of these problems.

It was easy enough to solve each symptom objection they raised but it was hard to convince them that the sound they expected to hear was not the sound that was recorded. They liked the hard, brittle, punchy, narrow sound because they were used to it, had never heard anything else. They had grown up with Hi-Fi not music. They were convinced that the emotion, excitement, thrill, etc., was in the sound of the music rather than being in the music itself.
It was only at the very end of the session, when I had solved all the problems that they finally understood what I had been saying. It was a cultural shock for them. I understood what they liked and why they liked it, but I had to get the whole system working right to show them they could have what they wanted and at a much higher level of enjoyment. They were an extreme case, they were experts

working in a Hi-Fi shop, they had no idea how much they didn't know! Credit where credit is due, they were attentive, respectful and listened with open minds and a willingness to learn. What more could I ask?

Problems vary from system to system and the varieties are infinite. I don't have all the answers, but I am working on it! Nor are everyone's expectations the same. A sound that makes nine out of ten people smile with pleasure may make the tenth person grind his teeth with pain. It's not for me to dictate the sound everyone should like. I can guide and educate but I respect your right to listen to a sound you enjoy, it is your system not mine, your music, your money.

My goal is to get as close to the music of a live performance as possible without any unnatural electronic Hi-Fi system artificiality. If you are seeking the same goal then this is the book for you!

The Invisible Problem that's Blighting Your System *see also page 33*

There's an invisible (and growing) problem that's blighting your system and preventing you from hearing it at its best. That problem is Radio Frequency Interference.

The Problem

Radio Frequency Interference (RFI) occurs when radio frequencies emitted by sources such as radio and TV broadcasts, mobile phones, electrical appliances and computers, affect the reception or operation of your Hi-Fi and Home Cinema equipment. RFI enters your system in two ways; through your mains supply, and by being picked up from the environment by your cables. And since there are so many more computers, wireless networks and mobile phones around these days, the bad news is that this problem is getting worse!

Any piece of wire connected to your tuner input picks up radio signals, so it is really no surprise that most cables act as aerials attracting airborne radio frequencies. All the cables in your system are a potential problem area, since in this context the radio frequencies they pick up become interference (RFI).

The Effect

It is difficult to convey in words just how destructive RFI is of high quality, musical sound in a Hi-Fi system. It gets into all parts of the system and degrades the sound in every way. RFI produces not only effects that I would call 'Hi-Fi' sound degradations (distortion, hardness, lack of deep bass, narrow dynamic range, etc.) but also 'musicality' degradation (destruction of rhythm and timing, flat dimensionality, narrow dynamic range, loss of low level ambience information, etc.). RFI is so pervasive that you probably won't even be aware of the detrimental effect it is having until you hear how good your system sounds when you have removed it! To find out more about why RFI is so much of a problem for an amplifier, see Appendix 1.

The Solution

Many manufacturers use various techniques to solve the RFI problem, but I believe that the unique woven cable design developed by Ray Kimber enables his cables to combat both sources of RFI (that being carried along mains cables, and airborne RF picked up by cables) most effectively.

The Kimber Kable weave technique, unlike some others, has no performance penalty and many advantages. It works by taking advantage of the 'capacitive cancellation' that occurs when the wires cross. The more crossings the more RFI cancellation, so the longer the cables are the greater the effectiveness. This cable design presents no coherent antennae pattern, so it resists the pick-up of airborne RF.

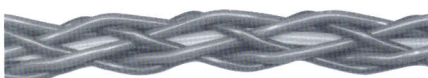

Diagram 1.4: Kimber Kable woven mains cable. Their interconnects and speaker cables also exhibit variations on this woven geometry.

RFI can be further combated during the manufacturing process by taking great care to ensure that there is a smooth high conductivity skin on each little copper strand. Since RFI travels primarily on the skin of a conductor (known as 'skin effect'), making the skin super conductive 'shorts' out the noise. VariStrand™ wire geometry (another Kimber development) - where each of the strands of wire in the conductors are of a different diameter - ensures that the 'skin effect' is spread evenly across the audio band, giving a neutral sound. Another bonus is that the increased conductivity lowers the dynamic impedance of the cable.

Fortunately there are a number of things you can do to combat RFI. Just replacing your mains cable with PowerKords™ should dramatically increase your system's performance since they are made using woven Kimber cable and reduce RFI on the mains.

Numerous customers have commented that their system sounded *"more than 10 times better"* after eradicating RFI by upgrading their mains, and changing their interconnects and speaker cables to KIMBER cables. Usually, though, people say that the improvement is so great they cannot put a figure on it, because the performance has changed as much in nature as in extent. It is not just better 'Hi-Fi', it is producing natural convincing music that it never produced before.

Clearly, keeping RFI out of your system is a major priority and I believe that Kimber cables are the key to helping you achieve this fundamentally important goal.

Chapter Two
The Power and the Glory

Upgrade Path Step 1: The Mains

I first wrote a guide to mains as a free infomation booklet entitled 'The Power & The Glory' back in 1998. It was a huge success (we've produced three editions over the years) and has been used extensively by both enthusiasts and electricians all over the country as a practical guide to the sonic effects of good wiring practice. Copies are held by the Radio Communications Agency in Wales, the reference library in Cardiff, and the reference library of the Royal Museum of Scotland in Edinburgh. For this chapter I have revised and updated the information in that original booklet.

This is a practical DIY guide that avoids unnecessary technicalities. If you want those, buy a few textbooks (check out the Further Reading section at the end of the book). The emphasis here is on practicalities not technicalities.

Though it is now generally accepted that mains quality can create problems for computers, electronics and electrical devices containing micro computers, and degrade quality on audio and video equipment, it is not generally understood why that should happen.

It has been assumed, in theory, that because the mains supply system is capable of supplying thousands of kilowatts of power at the nominal mains voltage of 230V 50Hz (in the UK), demand would not affect the quality of the supply. In practice, however, this is far from the truth. Not only does total demand come close to total capacity on a daily basis (generating capacity is closely and continuously monitored to match demand) but, more importantly, each user interacts with every other user.

How can this be? Well, of course, the supply is stiff like a solid object and Newton's law that 'Action Equals Reaction' applies in just the same way as in a mechanical system. If you take pulses of electricity from the mains (as the power supply in every piece of equipment does) you put equal pulses of demand on the supply. These demand pulses distort the 50Hz wave form. Some forms of demand pulse cause more distortion than others.

If everything affects everything else, what can be done to eliminate or minimise the effects? The answer is 'quite a lot'. Is it worth making the effort? Absolutely and most definitely yes! The degrading effect on the sound of a Hi-Fi system is enormous and the cost of the cure compared to the amount invested in Hi-Fi system hardware is a very small proportion.

My estimate is that poor mains quality reduces the sound quality and enjoyment of music from a Hi-Fi system by between 50% and 90%! I know that's a judgment call and other listeners may rate it differently but I believe mains quality is the major reason for dissatisfaction in Hi-Fi system sound quality. That is why I have devoted so much time and energy to finding the best cures for the many and various mains problems.

Investigating the Mains

When I first suspected that the electrical mains supply might be the cause of variable or unpredictable performance in a Hi-Fi system, questioning mains quality was considered heretical. Even six years later in 1980, an article I wrote for Hi-Fi Answers magazine brought letters of outrage that I should criticise the mains quality. These days, however, the mains is even suspected of causing cancer - oh, how times change!

My suspicions were first aroused by the observation that the sound of a Hi-Fi system varied with the time of day. Why, I wondered, did any system always sound at its best late at night, and its worst at meal times in the winter? There are, of course, many contributing reasons, but the first that I found, in 1974, was that the variation in mains voltage throughout the day matched the variation in sound quality of a Hi-Fi system. As demand for electricity increased the mains voltage fell and sound quality fell with it.

Testing my theory, I measured the output power and distortion of several amplifiers and found, as any engineer could predict, that reducing the voltage to an amplifier reduced the power output and increased distortion. A typical example was a Naim NAP 160, which delivered exactly half power with a mains voltage of 219 volts!

Tracking the mains voltage for a few days showed that 219 volts wasn't uncommon at all. Measurements taken in my own home revealed voltages as low as 203 volts on a Sunday lunch time! Whenever there was a high demand for electricity the voltage sagged. A very revealing observation was the gradual rise in voltage through the evening, reaching an all time maximum after about 11.30pm!

The case was proved, but this wasn't the end of the story by a long way. It was the start of a personal quest to solve mains problems affecting Hi-Fi sound quality that continues to this day.

Important: Safety and Legal Notes

This chapter describes the mains wiring system in use in the UK only. Wiring systems vary from country to country so check with your local codes and laws to make sure that you are not breaking the law or doing something incompatible with your local systems.

The techniques and advice given in this information book have been checked for accuracy and safety by Senior Technical Managers working for two different Electricity Boards. If your electrician presents any difficulties or disagrees with any of these techniques then he's not the man for you - find a better one!

What you can Work on

You must not interfere with either the main service cable (10), service head or cutout (8), meter (1), or meter leads between meter and service head. Your electrician can, however, work on the consumer unit (fuse box) (2).

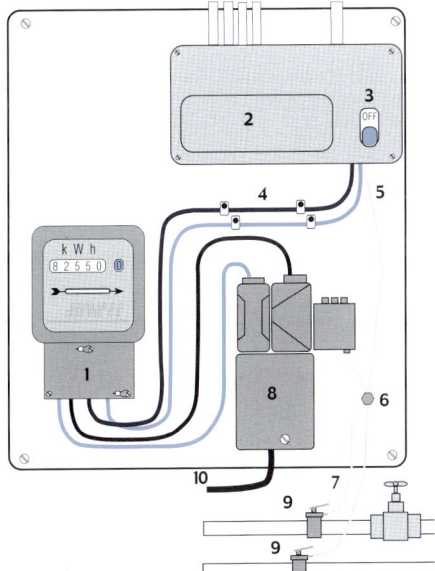

1 Meter
2 Consumer unit (fuse box)
3 Main isolating switch (on/off switch)
4 Meter leads
5 Earth cable
6 Consumer's earth terminal
7 Cross-bonding cable to gas/water pipes
8 Service head or cutout
9 Bonding clamps
10 Main service cable

Diagram from 'The Collins Complete DIY Manual', Jackson & Day, (1993).

Diagram 2.1: Main fuse board and consumer unit

What is Notifiable

Part P of the Building Regulations on electrical safety came into force on the 1st of January 2005. It states that *"anyone carrying out domestic electrical installation work must comply with Part P."* Where work is notifiable it must be certified by a *"certified competent person"* or the work must be notified to the Local Authority, which will then be responsible for inspecting and testing it for electrical safety.

Since then, Part P has been amended in an attempt to promote greater clarity of the requirement and to make enforcement more proportionate to the risk. To reflect these amendments, a new version of Approved Document P has been issued and it came into effect on 6th April 2006. [A copy may be obtained from the ODPM website www.odpm.gov.uk/electricalsafety.]

How does it apply to my mains upgrades? What is covered and what can you do without worrying about notification and fees?

Notifiable:
1. Installation of a dedicated Hi-Fi Ring Main
2. Fitting a dedicated Hi-Fi Consumer Unit

Not notifiable:
1. Earth bonding work
2. Fitting an earth spike
3. Extending an existing mains circuit
4. Replacing 13A outlet sockets
5. Fitting switch contact suppressors in light switches and central heating thermostats

Excellent, though rather technical, articles are available on the Institution of Engineers and Technology website www.theiet.org

How much Difference can Improving your Mains really make?

I have to say that the Hi-Fi Establishment was, for many years, quite dismissive of the importance of clean mains on sound quality. The attitude has been that as long as you connect 230 volts to your system, the sound quality you get depends on the quality of the equipment itself.

The truth is that poor mains quality ruins the sound of even the best equipment. In fact, the better the equipment the bigger the difference. On the other hand, even the most modest system benefits enormously from good mains.

An example of the importance of mains quality to sound quality is the effect that a high quality mains cable can have on even lowly portable stereos, as I described in the previous chapter (page 11). If you can hear that on the lowest of the low in Hi-Fi terms then just think how much bigger the difference will be on better equipment.

Mains quality is absolutely fundamental to the reproduction of music. I say 'fundamental' because the mains is the source of all the sounds from a system and the quality of the sound depends directly upon the quality of the mains.

My research has so far identified a number of problem areas, though not all of them have satisfactory solutions as yet. This chapter describes these problems and shares with you the solutions I have developed to overcome or minimise them.

Upgrading your Mains in 10 steps

The mains supply to your Hi-Fi system has several parts, but don't panic you don't have to do everything at once! Every improvement you make is worthwhile and worth doing in itself.

To get the best value for money and effort I have listed the sequence of steps below to reflect the importance and ease of attainment of the upgrades; complete all of these and you will cure all of the mains problems I have identified.

1. Leave your system on permanently (page 29)
2. Mains cables & distribution blocks (pages 34 & 48)
3. The earthing (page 38)
4. Harmonic noise filtering (page 35)
5. High voltage spike protection (page 37)
6. Contact cleaning (page 50)
7. The sockets (page 50)
8. Curing clicks, pops, buzzes & hums (page 46)
9. Wiring circuits (page 53)
10. The mains fuse board & consumer unit (page 49)

Two Easy and Free Mains Upgrades

1. Leaving your system on permanently

One of the simplest mains upgrades you can do doesn't require any work. Simply leave your equipment switched on 24 hours a day, 7 days a week. (Do unplug equipment from the mains during electrical storms, remembering to also disconnect TV and FM aerial leads!)

The sound from your system will take at least a week to settle down and stop changing. If, subsequently, you turn any part of it off, even for a few minutes, expect it to need 24 hours to settle again.

So what are the benefits of leaving your system on permanently? The improvements in smoothness, sweetness and musicality are enormous. Bass may appear to drop half an octave and improve in clarity. The whole sound will be more relaxed and 'listenable'. It will also reduce the incidence of breakdown. Switching electronics on puts enormous stress on the internal components and this shock damage is the main reason for component failure.

The average system burns a few watts 'idling' and a few tens of watts when playing so the cost in electricity is really very low. By my reckoning the average system component only costs around 10 pence a week if left on permanently. Even if you have a power hungry system it is unlikely to cost you more than a couple of pounds a week. By comparison, a 100W light bulb costs £1.38 per week if left on permanently, and a washing machine (3700W) used once a day costs £2.12 per week (based on electricity rate of 8.2p per unit). A customer once told me that leaving the system on 24 hours a day was the biggest and cheapest upgrade he had ever made. A few pence a week for 'the biggest upgrade' he had done? Sounds like a bargain to me! For more information on costs and offsetting your carbon, see the **Appendix III** at the back.

Some enthusiasts have been concerned about safety and the potential fire hazards. Amplifiers shouldn't overheat, and unless faulty or badly designed will reach a steady working temperature and then get no hotter. Some amps, both transistor and valve types, are designed to run quite hot.* Our enquiries to RoSPA have revealed no recorded incidents of a Hi-Fi system causing a house fire. Even if a component goes 'on fire' itself, it is self-extinguishing and there is nothing inside a piece of Hi-Fi to sustain a fire (i.e. there is nothing to catch fire).

Safety Note: Valve Equipment. It seems that an alarming proportion of modern valve power amps are not sufficiently well designed to make 24 hour a day use safe. Typically they lack robustness and protection in the event of valve failure leading to considerable 'collateral damage'.

To make matters worse modern valves do not enjoy the reliability of the old ones. The switch-on surge is the highest stress time for a valve so switching a valve power amp on and off every day will increase the fault rate, but at least you can immediately switch it off again if a fault occurs. The best compromise is to switch your power amp on an hour or two before you want to listen to your system. Valve preamps do not seem to share this problem so can be left on permanently.

2. Switch off your Lights!

I have been aware of the degrading effect of different types of lighting for many years but have found no type of electric light that didn't have an impact on sound quality. Large fluorescent lights are probably the worst culprits because they rattle so much in response to the music and also generate large amounts of RFI onto the mains and directly into the 'ether' (now called the 'airwaves'). Ordinary incandescent lamps (40, 60, 100 Watt light bulbs) audibly degrade the sound because they ring and vibrate in response to the music in a very audible way. The new and popular low consumption lamps are probably equally degrading as they are high frequency fluorescent and produce RFI.

Many people have reported that their systems sound better in the dark and as I've just explained there is a physical rather than psycho-acoustic reason why. Try candles, they have no degrading effect and induce a more relaxed ambience into listening. Just be aware of the potential fire hazard!

Voltage and Frequency Variations

Your Electricity Board is obliged by law to deliver 230 volts +10% - 6% (i.e. between 216.2 volts and 253 volts), and to maintain the frequency at 50Hz ± 1% (i.e. between 49Hz and 51Hz) over a 24 hour period.

The EU decided, in its wisdom, to harmonise the UK standard mains voltage of 240V AC and the European standard of 220V AC, at 230V AC. This sounded fine in theory but when the cost of replacing all the supply equipment to deliver 230V was calculated it was decided that it was uneconomic (there being no advantage whatever in changing, other than 'harmonisation'). However, to avoid accusations of failure to harmonise, they simply fiddled with the legal voltage limits.

The law now states 230V +10% -6%, thereby allowing the European 220V system to stay at 220V and the UK to stay at 240V, yet both appear to be harmonised!

Mains Voltage

Your house is supplied from a transformer at a sub-station (serving 10-100 homes) and is what's called single phase (one Live, one Neutral). The substation is supplied with three phase 11,000 volts and distributes the load to customers by balancing the demand across the three phases.

As new consumers are added to a phase or the existing consumers use more power, that phase gets overloaded and the voltage and frequency sag below the permitted limit, particularly at times of high demand like meal times and TV programme breaks. The effect of 20 million kettles being switched on at the same time can stretch the National Grid supply to breaking point!

The voltage of the '11,000 volt system' is automatically regulated at the Primary substations to about 11,200 volts, and is maintained at this value over a wide range of system loadings. Most of the voltage problems come from the 230 volt domestic system which becomes overloaded over time. However, the supply companies insist that most customers have a supply within the statutory limits all the time, with a few exceptions.

Low voltage problems can be caused by poor internal house wiring, or by an inadequate sub mains to flats and outbuildings, as well as by overloading of the substation.

At the other end of the scale, don't ignore the possibility of over voltage. It costs a fortune in light bulbs, causes an increase in distortion in Hi-Fi systems and reduces reliability.

The Effect of Incorrect Mains Voltages

As I described earlier, the mains voltage has a profound effect on the power output and distortion of a power amp. The mains voltage also has a surprising effect on the performance of a CD player. The motor speed is voltage controlled, not frequency controlled with a crystal reference as most people would expect.

Even though the motor supply voltage is regulated, the mains voltage does affect it, making it play slightly faster or slower.

The effect on the music is quite large, making it sound either fast and exciting or slow and relaxed. I discovered this whilst experimenting to find out why a 'Pandora' isolating transformer, that was claimed to improve CD sound quality, did affect the sound. I discovered that the 'Pandora' was simply increasing the mains voltage by 24 volts, hence increasing the play speed and making the sound appear more 'exciting'. Conversely, low voltages can more than halve your amplifier's power output and massively increase distortion.

If you suspect that your mains voltage is sagging below the permitted volts limit contact your Electricity Board and ask them to check it by monitoring it over a 24 hour period. They will do this free of charge (but they may need a firm hand as the cost to them of correcting your voltage could be very high).

I don't know what chart recorders they use these days or how they are calibrated, but when I complained of low voltage (203 volts on a Sunday lunchtime!) in 1973 the Board engineer calibrated the recorder in my house with a simple multimeter that measured 10 volts high! After a short argument he set the recorder correctly with my calibrated digital voltmeter.

Testing the Effect of Voltage Variations

To test the effect of mains voltage variation, Paul Houlden, a transformer manufacturer, made a special transformer for me with voltage tappings at 0.9, full mains, and 1.1 times mains voltage.

These small changes were enough to change the tempo of the music on a CD by clearly observable amounts. Paul plays drums, so the changes in tempo were very obvious to him. We estimated that each '0.1' voltage change in either direction caused a 1% speed change. I must emphasise that these were subjective listening assessments as we were unable to measure the speed of the CD itself.

Mains Frequency

Power demand in the electricity supply reduces the mains frequency as well as its voltage. As the legal requirement is that mains frequency must be maintained at 50Hz over a 24 hour period, suppliers are able to increase the frequency at certain times to compensate for drops at other times. The frequency is commonly increased late at night.

These fluctuations in mains frequency have an effect on a Hi-Fi system. Most turntables use AC synchronous motors whose speed of rotation is locked to the mains frequency. A turntable with a synchronous motor plays slightly slow when the frequency drops at times of high demand and slightly fast when the frequency is higher at times of low demand (e.g. late at night).

When it plays slow, the music is slightly down in pitch and dull; when it plays fast the music is pitched higher and sounds fast and exciting. In my experience many turntables are set by the manufacturer to play slightly fast (up to 1.5%). I assume that they do that to compensate for low mains frequency and because it makes their product sound more dynamic.

Recently some manufacturers have produced regenerative mains supplies that include the facility to increase the mains frequency, claiming that this can improve the sound quality of the system. I suppose it depends on what you call an improvement!

Radio Frequency Interference and the mains supply

The house wiring maintenance and upgrading steps described in this chapter are essential to good sound quality but RFI on the mains is a separate problem. It is a seriously degrading pollutant that needs to be dealt with however good or bad your house mains wiring.

Radio frequency is the frequency of radio waves that are between about 10 kilohertz and 0.1 terahertz. Radio waves are all around us in the environment being generated by a multitude of sources:

• Radio and TV broadcast stations worldwide

- Two-way radio transmitters (Citizens Band, emergency services, taxis)
- Mobile phones and pagers
- Household electrical appliances (e.g. fridges, freezers, washing machines, vacuum cleaners, toasters, power tools)
- Computers and any equipment containing microprocessors (e.g. CD and DVD players, video recorders, TVs), and wireless networks
- Business (e.g. cash registers, power tools, industrial machinery)
- Fluorescent lights and light dimmers (see also **Clicks and Pops**, page 46)

Radio Frequency Interference (RFI) is when these radio frequencies affect the reception or operation of other equipment such as your Hi-Fi system.

Most cables act as aerials attracting airborne radio frequencies, so all the cables in your Hi-Fi and Home Cinema system are a potential problem area. And don't forget that RFI has had plenty of opportunity to enter the mains cable carrying your mains from the power station all the way to your home.

RFI affects sound quality in a number of ways such as loss of rhythm and timing, flat sound lacking in dimensionality, narrow dynamic range, loss of information, distortion, hardness, and lack of deep bass. RFI is so pervasive that you probably won't even be aware of the detrimental effect it is having until you hear how good your system sounds without RFI. To find out why RFI is so much of a problem for amplifiers, see Appendix I, at the back of this book.

Removing the RFI carried on the mains before it enters your Hi-Fi system, and then preventing further pick up of RFI by cabling within your system, is clearly fundamentally important to achieving great sound. The result is a clearer, cleaner, more natural and musical sound. But how do you achieve this? Some Hi-Fi mains cable companies use filtering techniques to reduce RFI getting from the cable into equipment. But most ignore this problem altogether.

So how do you Remove RFI?

The answer is to use cables that remove RFI from the mains supply feeding your system and reject further RFI pickup. Those cable manufacturers that do address this problem use a variety of techniques, but I believe the woven cables made by Kimber Kable achieve this in the most effective way and are the very best available.

Whichever manufacturer's cables you use to solve the RFI problem, to make the job more manageable I suggest that you upgrade your cabling in stages. To get the most improvement for your money start at your equipment and work back towards your consumer unit.

Of course it may not be possible to replace all the mains cables or you may wish to do it in stages. In my experience, the order of importance for replacing mains cables (in terms of the degree of improvement you will gain) is:

1. **CD player** (or CD transport if you have a two box CD transport and DAC), **or your primary music source**. CD and DVD players are the most sensitive to RFI on the mains and therefore show the biggest improvement.
2. **DAC** (if you have a two box CD transport and DAC)
3. **Preamp**
4. **Power amp**
5. **Tuner** (and any other music source)

It's worth also noting that the ordinary plastic extension blocks available in shops degrade the sound to a surprising extent. Factors such as the build quality, the materials used (some aren't even fire proof!) and the design of the cable attached will all have an impact on the quality of sound, for example, by acting as an aerial for RFI, or having poor contacts that increase impedance (see **High Impedence**, page 48).

To reduce the impact on your Hi-Fi of interference put onto the mains by your household appliances, install a dedicated Hi-Fi ring main from your consumer unit. (For details on how to do this, and discussion of ring versus spur circuits, see **Mains Circuits**, page 52). Using a cable which cancels RFI already carried on the mains, and which repels airborne RFI, is clearly desirable and can make a significant difference over this distance. I highly recommend our Ring Main cable made by Kimber Kable - expensive, but it has very low impedance and its RFI cancelling abilities are incredible. The effect on a system is enormous: much deeper, cleaner bass; faster and more natural sound; and sweeter treble.

Once you've removed RFI from the mains, you naturally don't want to dilute this improvement by allowing RFI to get back in. Using interconnect and speaker cabling that cancels RFI rather than attracting it makes complete sense. Again, thanks to their woven construction, I highly recommend Kimber Kable's range of interconnects and speaker cable which have been uniquely developed for this capability.

Mains Noise and Harmonics

Mains noise and harmonics have an insidiously degrading effect on the sound of a Hi-Fi system (or indeed the sound and picture quality in a Home Cinema system). They add a grainy 'mush' quality to the sound that hides detail and spatial resolution. They are another of those mains problems that vary widely from time to time during the day, causing your Hi-Fi to sound better at some times than it does at others.

Where does Mains Noise come from?

Much of the noise and harmonics plaguing your system is generated within your home. Anything connected to the mains can put noise and harmonics back onto the mains. Try this experiment. Go round your home and unplug everything that's connected to the mains - everything, including your fridges and freezers (though don't forget to plug them back in afterwards!). Now listen to your system. The chances are that it'll sound clearer and more musical.

Components in your Hi-Fi or Home Cinema system itself can also put noise onto the mains and degrade the sound; power amplifiers and digital components are common culprits. Noise problems from outside your home come from power lines, radio and TV stations, industrial plants and factories and, of course, your neighbours!

There are devices available that let you 'hear' the mains noise which are great if you want to find the sources of mains noise in your home, decide where to best place a mains filter and/or assess the effectiveness of a mains filter. Audio Prism make one called the Noise Sniffer™, which is available on loan from www.russandrews.com for a small charge to cover delivery and collection.

Solving the Problem of Mains Noise

Many companies make mains conditioners that claim to solve the problem. Over the years I have auditioned dozens of devices described as 'mains conditioners' designed to be connected between your Hi-Fi system and the mains. These devices vary from simple to very complex inductive/capacitive filters. They break the mains supply path, inserting inductors and capacitors in series with and in parallel across the mains. The result is a resonant filter that does indeed remove some noise but regrettably does not deal with all frequencies equally across the spectrum.

The typical results are frequency response graphs like those below.

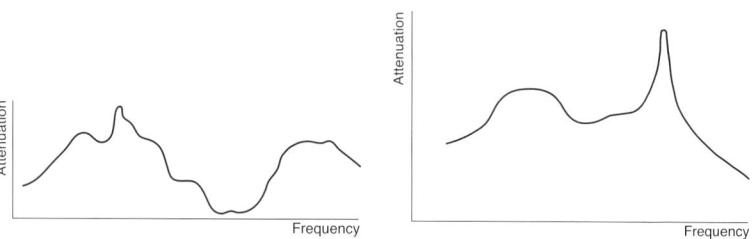

Diagram 2.2: Graphs showing typical frequency response of mains filters

(Reprinted from 'Radio Interference Supression' by GL Stephens, pub. Lliffe & Sons, 2nd Edition 1952.)

The frequency response of such filters imposes itself on the frequency response of the system supplied from it. Just as bad is the high dynamic impedance presented by such filters. This slows down the resupply speed of the system power supplies, degrading and/or removing musical elements such as 'life', dynamics, bass and treble.

My research has shown that the 'inductive resonant' filters used in most commercial and industrial 'mains filters' have a degrading effect on sound. Although they do suppress some of the RFI, the inductor and capacitor resonant circuits raise the dynamic impedance of the mains to unacceptable levels. The lower the mains impedance, the better the power supply in the equipment works and the better the sound quality (see **High Impedance**, page 48).

I devoted 3 years research and development to developing a solution to this problem and made two key findings. First, that by using 'passive differential' and 'common mode' filtering it is possible to produce outstanding noise reduction without degrading side effects. Second, that by designing them to work in parallel with your system, rather than drawing the power through them to your system, you can ensure that the mains impedance is not increased. My own Silencers and Purifier mains conditioners are, of course, designed to work like this. However, whichever mains filter you choose, ensure it is a parallel (not a series) type.

High Voltage Spikes

High voltage mains spikes come from a variety of sources such as lightning, fluorescent lights, refrigerators and washing machines and are a serious source of distortion in equipment. Because the quality and severity of the spike activity varies during the day, and from day to day, the sound of a system will vary and the effects will be greater at some times than at others. The average electricity supply receives transients of up to 10,000V.

Although the spikes are inaudible, their effect is not. You only become aware of the degradation they cause when they have been removed. Solid state equipment, semiconductors and switching devices are particularly vulnerable to such surges and spikes.

The high voltage spikes cause audible degradation of Hi-Fi sound by their effect on the power supply of a piece of equipment and by raising the background noise level in amplification circuits. The spikes saturate the core of the mains transformer, preventing normal transformer operation during the period of each spike and causing distortion to the waveform.

Hysteresis bounce magnifies the voltage of the spike so that it appears even bigger on the transformer's secondary winding, appearing on the output of the rectifier as noise. It can be seen then that mains spikes reduce the capacity of the power supply by 'strangling' the transformer operation, producing effects on a power amplifier like softened bass and increased high frequency distortion.

The spikes also cause premature ageing and failure in sensitive components like transistors, diodes, microprocessors and integrated circuits. Cleaning high voltage spikes up will, therefore, give your electronic components longer life and lower repair bills.

In computers the transient mains spikes cause noise that is very similar to data pulses. They are interpreted as corrupt data or switching instructions and so cause random 'lock-outs', freezes and unintended operational errors. It's like a constant blizzard of false data and instructions that randomly affect your work depending on what operations the computer is performing when they occur. Often it looks like operator error when, in fact, it was an ill-timed spike.

I have been making and selling spike suppressors for many years. In the early Eighties I found very fast acting components culled from Aerospace technology that would absorb the energy in the spikes without introducing any impedance changes as they operate across live and neutral. More recently I discovered a component used for surge protection in the mobile phone industry. By using them, you hear an improvement in the bass depth and clarity in your system, plus the sound is smoother and cleaner, more detailed and dynamic. The benefit of both devices is that their presence is absolutely undetectable in the absence of high voltage spikes.

Poor & Noisy Earth

Upgrading your house and system earthing is very worthwhile. Providing a good clean earth for a Hi-Fi system can give a cleaner, deeper bass and a more stable soundstage. In addition, much of the noise problem commonly experienced in Hi-Fi systems is caused by poor earthing. *House Earthing* describes your mains supply earthing, and *System Earthing* describes your Hi-Fi's connection to the house earth.

Alternative ways your home will be earthed

There are four ways that your property might be earthed, and these are listed below:

1. Earth Bonding Clamps to Water Main Pipe

If you live in an older property it will probably be relying on earthing to the incoming cold water supply pipe using earth bonding clamps (see Diagram 2.3 below). However, if the earth bonding clamps are inadequate or dirty then instead of being an effective 'earth' the earthing and cross-bonding becomes an 'aerial' for noise, spikes and RFI. If your water main pipes and connections are non-metallic, non-conductive plastic ones, a different earthing method is needed, using an earth spike. In a properly earthed system the gas and water supply pipes and any other exposed metalwork are bonded to earth. Check that your system has all these main and supplementary bonding conductors; install any that are missing.

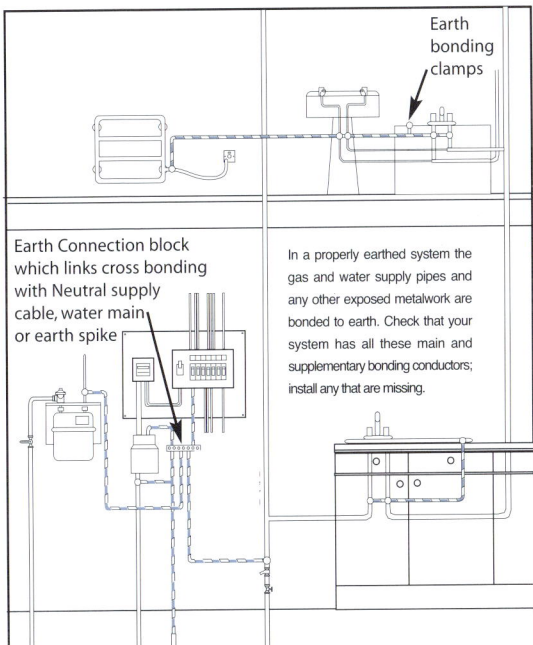

Earth bonding clamps

Earth Connection block which links cross bonding with Neutral supply cable, water main or earth spike

In a properly earthed system the gas and water supply pipes and any other exposed metalwork are bonded to earth. Check that your system has all these main and supplementary bonding conductors; install any that are missing.

Diagram 2.3: House earthing showing earth bonding clamps to water main pipe

2. Earth Spike (Earth Electrode) from an Earth Connection Block

Sometimes the house is earthed via an earth spike (a 4ft copper or copper clad steel rod) that is connected to the earth connection block and placed in the ground (outside or in the basement). You may discover that your home has this in addition to one of the other earthing methods.

3. Connection from a Cut Out Block to an Earth Connection Block

In most urban houses a connection is provided from inside the Electricity Board's cut-out block to an earth connection block. This provides an effective path to earth along the metal sheath of the main service cable to the substation where it is connected to earth.

4. PME (Protective Multiple Earth)

Many modern homes, and those in country areas, are provided with a kind of earthing called PME. The earth current is fed back to the electricity substation along the neutral return wire, connected to earth at frequent intervals along its length.

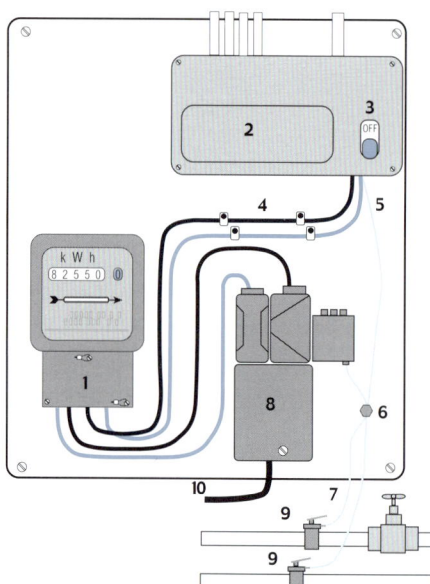

1 Meter
2 Consumer unit (fuse box)
3 Main isolating switch (on/off switch)
4 Meter leads
5 Earth cable
6 Consumer's earth terminal
7 Cross-bonding cable to gas/water pipes
8 Service head or cutout
9 Bonding clamps
10 Main service cable

Diagram from 'The Collins Complete DIY Manual', Jackson & Day, (1993).

Diagram 2.4: Main fuse board and consumer unit

House Earthing

Do check the earthing you have in your home, it is there to protect you from an electric shock in the event that an appliance should go faulty. A number of customers have discovered to their horror (and mine) that they had no cross bonding at all in their house wiring systems. Even if you have no apparent problems, it is wise to check all the cross bonding connections between water pipes, etc. You will find them attached to all exposed pipe work beneath sinks, baths, etc. They get loose in time, and dirt and corrosion ruin the earth contact. Slacken each clamp, slide it along the pipe line an inch or so, clean the place it was on with wire wool and a good contact enhancer (I always use DeoxIT™), move it back and re-tighten.

If you don't already have one, you can improve your general house earthing by adding an earth spike from the earth connection block (as described in point two page 39). Improving your Hi-Fi system earthing by the addition of one or more earth spikes connected as close to your Hi-Fi as practicable is also very beneficial. (See instructions on the next page.)

Using a PowerKord™ or other quality mains cable can highlight an earthing problem:

Mr Bolan of Montrose phoned us to pass on an experience that may be of benefit to many of you. He has a Naim system and was concerned about a distortion problem with the amplifier. He experimented with our PowerKords™ and found that, if anything, the problem was made worse. He took his amp to his dealer for testing and received a clean bill of health; his Electricity Board fitted a mains monitor without finding any fault. It was only when his washing machine went faulty and was pulled out for repair that it was found that the earth bonding straps on the water pipes behind it were loose. Cleaning and tightening them cured the distortion problem with his amplifier!

This is a good example of the importance of distinguishing between a symptom and a cause. It would have been easy for Mr Bolan to reject the PowerKord™ believing it made his system sound worse rather than highlighting an existing problem.

Earth bonding was also to blame for this mysterious problem:

I once had a bizarre problem to solve where the Hi-Fi system crackled when the downstairs loo was flushed! I advised the bemused customer to dig up round the water mains to see what it was made of. Two feet out from the house was a plastic connector between two metal pipes. A cross bonding strap from one pipe to the other cured the problem and improved the sound of the Hi-Fi system.

It is worth mentioning that during the summer months the ground dries and so earth impedance goes up. It is a common cause of poor sound quality in the summer.

Fitting an additional Earth Spike

When burying the earth spike, choose a place a little away from the house where the soil will be moist - and avoid buried pipes and cables! Be sure to keep the ground round the spike well watered to maintain a good low resistance earth. Several spikes may be used in a star pattern, called 'external star grounding' to increase the quality of the earth. If you find it difficult to bury your earth rod (they sometimes bend when they are driven into the ground), you may find that a better solution is to use a 2m length of pure copper 22mm diameter plumbers tubing. Dig a trench about 1m deep (this can be done quite neatly under your lawn) and lay the tube into the bottom of the trench, horizontally. Connect the earth cable that's connected to your system to the copper tube – you will need to fabricate a strap, or you could simply drill a hole in the end and bolt the cable on. Re-bury the tube and you'll have an effective, low resistance earth for your system.

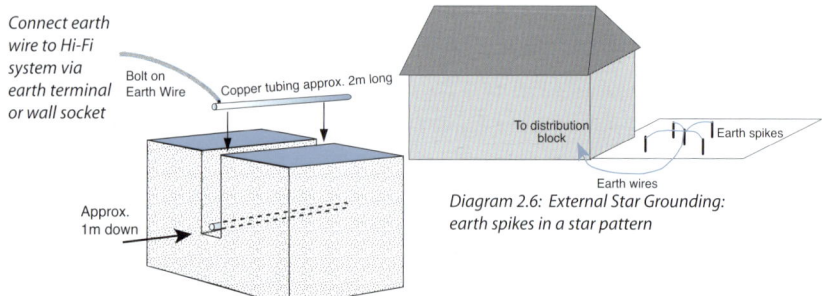

Connect earth wire to Hi-Fi system via earth terminal or wall socket

Bolt on Earth Wire

Copper tubing approx. 2m long

To distribution block

Earth wires

Earth spikes

Diagram 2.6: External Star Grounding: earth spikes in a star pattern

Approx. 1m down

Diagram 2.5: Showing how to bury an earth spike horizontally

The earth wire should be at least 4mm^2 and preferably 10mm^2. Alternative low impedance earth wires are available: see Ben Duncan, 'Ground Work', (Hi-Fi News March 1997), and the Kimber woven earth wire. Customer, Mark Rose, contacted me to suggest sealing the final contact between the earth cable and the copper tube with a whole tube of clear silicon sealant, left for 24 hours to fully dry inside and out. I think this is good practice, though I actually leave mine so that I can periodically check and retighten the bolt.

The earth wire from the earth spike should be connected as close to the Hi-Fi system as possible. It is best connected to your Hi-Fi distribution block (some good mains blocks feature an earth terminal for this purpose), but you could connect it to the wall socket serving your system. Feed the earth wire behind the wall socket and connect it to the earth terminal in the back of the socket along with the ring main earth wires (or use my SuperSocket with earth terminal on the front to make the job much easier!)

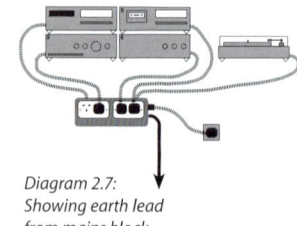

Diagram 2.7: Showing earth lead from mains block

System Earthing

Whilst it is true that all items in a system are interlinked via the signal grounds on the interconnect cables, this is not the same as good chassis or 'star' grounding. Confusing chassis earthing and signal earthing will degrade bass performance and often causes hum.

Chassis earthing is where your system case work connects to the mains earth. Signal grounds are usually connected to chassis grounds via a resistor to prevent hum induction into the signal.

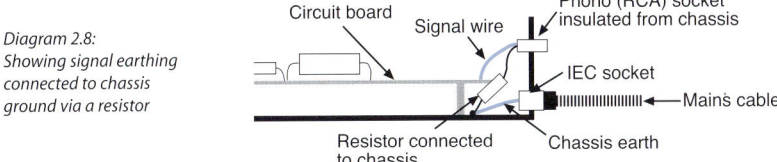

Diagram 2.8:
Showing signal earthing connected to chassis ground via a resistor

Circuit board
Signal wire
Phono (RCA) socket insulated from chassis
IEC socket
Mains cable
Resistor connected to chassis
Chassis earth

You can get improvement in the sound of your system if you improve the grounding (earthing) of your metal-cased equipment. One of the reasons is because hum is induced into equipment casework by its mains transformer, and earthing it can remove some of the hum and improve your equipment's performance. In principle all the hardware chassis (casework) should be connected independently to one main earth point, see below:

Improve your chassis (casework) earthing

It's simple to do - simply slacken off one of the screws that hold your casework together, wrap a piece of earth wire behind it, and attach the other end to an earthing point. You could simply connect the earth wire into the big earth pin of a mains plug (only connect to the earth pin!) and plug that into a wall socket. Some loudspeakers - for example, Tannoy and Russ Andrews Quave LS1 speakers - helpfully provide an extra binding post that lets you earth the speaker drive units for better sound quality. Don't connect your earth wire to a turntable grounding point on your amplifier - that's a signal earth and not a chassis earth.

A Star Grounding Block comes into its own when you have more than one casework to earth. You get better performance if you connect all of your casework earth cables to one earthing point - known as 'star earthing' or 'star grounding'. A Star Grounding Block lets you connect the earths from several pieces of equipment - tidily - to one earthing point. It means you don't have to try to cram eight pieces of earth cable into a mains plug.

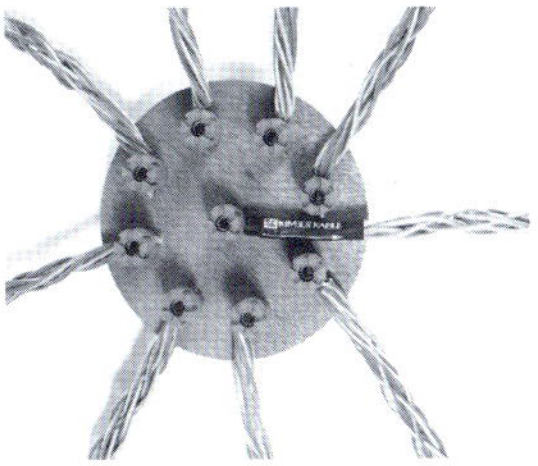

earth cables from equipment casework

earth cable to earth point

Diagram 2.9:
Star Grounding Block in use

Occasionally, when a system is connected together, a hum is heard coming from the speakers. This is often described as an 'earth loop'. The usual advice is to remove the earth wire in the mains plug of the offending piece of hardware. This solution may break the 'loop' and silence the hum but it is usually only hiding the symptom not curing the cause.

One of the most common causes of the problem is hum induced into the chassis of the equipment by its mains transformer (see the description of Electromagnetic Fields below). If your equipment has only two wire connection (i.e. no earth wire) you may gain an improvement in performance by connecting an earth to a chassis screw. Sometimes this adds a 'buzz' into the sound but if it doesn't then you will get more 'solid', stable sound and better bass. Experiment!

Electromagnetic Fields (EMF)

Electromagnetic fields are generated by the voltage in a cable of anything powered by AC electricity. The electromagnetic field radiates from a wire rather like a magnetic field radiates from a magnet.

The electromagnetic field is of interest for two reasons. First, the field generated by the power supply in any item of Hi-Fi equipment significantly degrades the sound quality by polluting nearby circuits. The effect of removing it is clear and obvious to anyone. It has been compared to the difference between a £300 CD player and one costing £1,000!

Second, there is a growing body of evidence being accumulated in the UK and in the USA that strong magnetic fields are hazardous to health. The main concern here is the very strong field radiated from TV and computer monitor screens. Any method of reducing the field strength will reduce the hazard; an important consideration for anyone working at a computer every day. (To date, no link has been established between EMF and cancer - but research in the USA and the UK is continuing on this subject.)

The EMF field from the mains transformer is made bigger or smaller depending on which end of the primary winding is connected to Live or Neutral (see below).

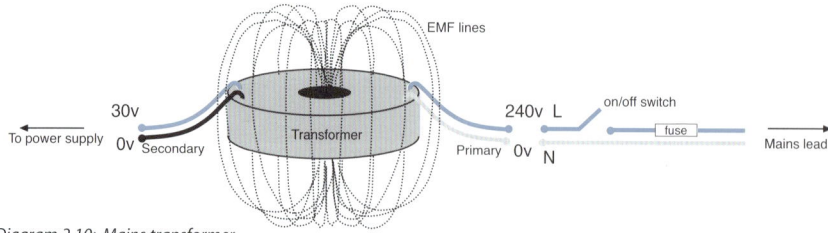

Diagram 2.10: Mains transformer with Electromagnetic Field

The strength of the electromagnetic field can be minimised by connecting the mains polarity correctly. The field is up to 1,000 times stronger if the polarity is wrong. Unfortunately, connecting the correct colour wire to the Live pin in the mains plug will not guarantee minimum field, because everything depends on how the mains transformer is connected up in the equipment.

- 0V Neutral or 240V Live

Secondary | Primary

- 240V Live or 0V Neutral

Diagram 2.11: Transformer primary and secondary windings

From a safety standpoint, it doesn't matter which way round the Live and Neutral wires are connected to Hi-Fi equipment because neither Live nor Neutral are connected to the chassis (or anything else except the mains fuse and on-off switch contacts). In any case, some equipment (like CD players) is supplied with detachable leads and reversible (Figure 8) plugs. Nor does it matter electrically. Some manufacturers connect Live to the 'top' of the transformer winding, whilst others connect it to the 'bottom'. Both ways work but one way creates the smallest electromagnetic field. The way that creates the smallest field sounds noticeably better than the way that creates the greatest field.

Correcting equipment polarity

Since there is a noticeable improvement in sound and picture quality when the polarity is correctly connected, it makes sense to try and get it correct. You can do this by eye and ear... simply reversing the way that Live and Neutral from the mains lead enters the equipment. If the lead has a Figure 8 plug simply turn this over. Otherwise just reverse the Live and Neutral wires in the 13A mains plug (see Diagram 2.12 below) - yes, this is safe to do!

One way will sound more natural and more musical than the other - and there may also be less transformer 'hum'. This is the correct way! If you are testing a number of pieces of equipment, prepare a cable with Live and Neutral reversed. Then you can compare in turn the performance from your equipment with a normal cable and the one where you've reversed Live and Neutral .

SAFETY NOTE: This advice applies only to Hi-Fi, audio, TV and video equipment. DO NOT try it on ordinary domestic appliances, unless they are double insulated, as there may be a hazard on some older types.

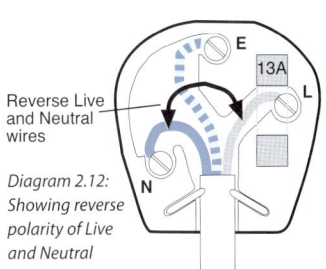

Reverse Live and Neutral wires

Diagram 2.12: Showing reverse polarity of Live and Neutral

Clicks, Pops, Buzzes and Hums

Many of you will be all too familiar with the problem of mysterious noises heard through or from your Hi-Fi! These noises fall into several common categories: clicks, pops, buzzes and hums. Here I'll explore the causes of the noises and advise on steps you can take to cure or minimise the problem. (You may find that implementing the checks and upgrades discussed elsewhere in this book will remove the problem without further attention.)

Clicks and Pops

Clicks and pops are heard through your speakers. They are commonly caused by thermostats (e.g. in fridges, freezers), and by the starters in fluorescent lights arcing as they switch on and off. The switch arcs as it operates and not only puts a 'click' onto the mains but also puts out a radio frequency pulse which can be picked up by Hi-Fi leads and be amplified as an audio signal. It is this RF 'click' which is so annoying in a Hi-Fi system and, of course, cannot be removed by mains filtering. The RF 'click' can only be prevented by suppressing the switch that causes it. (You may also hear clicks and pops caused by your neighbour's household appliances).

The first step is to check that the rest of the house wiring is in good condition and without fault. A loose screw in the Neutral connection of a mains outlet or hidden junction box can cause the whole ring circuit to act as a big transmitting aerial for switching clicks. Junction boxes are hidden under floors - so an electrician may be needed to trace them and cure the fault.

Once you've ensured that your wiring is electrically sound, you may find that the problem can only be cured at the source. Identify the offending switches or appliances and put a switch contact suppressor and VDR across each switch contact terminals. In the case of fluorescent lights, the only cure is to replace the neon flashover starter normally fitted with an electronic starter.

Buzzes and Hums

If the buzz or hum is mechanical in nature you will hear it from the equipment, not through the speakers, and it will often fluctuate in level at different times of day. Common causes include DC on the mains, RFI, or the transformers in your equipment.

The transformers can be at fault either due to the size of the EMF they produce or simply because their design (or economies in their manufacture) makes them susceptible to fluctuations in voltage with the result that they become noisy.

Alternatively, the noise may be constant or cyclical in nature, and heard through the speakers. There are a multitude of causes for this, the most common are, poor earthing or an earth loop problem, mains noise, and RFI (from household appliances, from household ring wiring running past RFI sources like fluorescent lights, or from outside your house, e.g. power lines, radio stations, factories, your neighbours!)

A combination of one or all of the solutions below should cure or reduce your buzzes and hums, but ultimately you may just have to live with them (or change your equipment).

How to cure Buzzes and Hums

Ensure your mains connections are clean and tight. Check that you have earth bonding and that the clamps are clean and tight. Also consider fitting an additional earth spike to your Hi-Fi via an earth terminal. See **Poor & Noisy Earth** on page 38 for full details on solving an earth loop problem.

I have found that fitting a PowerKord™ can reduce the noise by up to 50%. Their inherent RFI rejection and filtering properties greatly reduce the noise polluting your system. I'd also recommend a good parallel noise filter; often several are needed to create a 'quiet' mains environment.

Check the polarity of your equipment to ensure that the electromagnetic field is minimised. Don't stack your equipment directly on top of each other as EMF generated by the transformers will pollute the circuits of the equipment above and below.

Transformer noise due to fluctuations in voltage is a difficult problem. You can do nothing about the voltage fluctuations, so you are left with either making the noise less intrusive or changing your equipment.

You can minimise the noise vibration by damping the case of your equipment internally. Sound Deadening Pads and Damping Pads are ideal for this. You may need to remove the bolt securing the transformer and put a rubber pad between the transformer and the base of the equipment casing.

Try fitting our Oak Cone Feet to help dissipate the vibration from the casework. Avoid metal equipment racks, as they can track the hum all round the system, instead choose wooden equipment racks.

High Impedance

Impedance is the resistance to an alternating current. Although Hi-Fi and Home Cinema systems have relatively low current demands, maintaining a low impedance is important to ensure good bass performance and treble sweetness.

Low impedance is needed to allow an amplifier to draw the transient current it needs to accurately follow fast, current hungry music transients. A modest Hi-Fi amplifier can easily demand 100 amps from the mains for a milli-second on loud drum transients! Lowering the impedance also reduces voltage drop in the supply.

If the circuit that your Hi-Fi is plugged into has too high an impedance, the result will be a flat, two-dimensional soundstage with poor bass and treble. High impedance is caused by inadequate mains cabling and dirty or loose contacts from your equipment leads back to your consumer unit. The use of fuses, circuit breakers and residual current devices will also have an impact on impedance.

Mains Cables and Extensions

The impedance of a cable is determined by the thickness of the copper wires, the number of wires and the purity of the copper. Increase any of these factors and you will reduce the impedance of the cable. Reducing the impedance of the cable means you increase its current handling capacity, allowing more current to flow. Basically, if you double the size of any given cable, you will halve its impedance.

Simply replacing your equipment's mains leads with higher quality (larger diameter) Hi-Fi mains leads will reduce mains impedance. Russ Andrews mains cables are all low impedance cables (as well as RFI cancelling). In addition, the type and quality of your mains extension block will affect mains impedance. A high quality extension block will include high performance, high pressure sockets and high quality cabling to keep mains impedance to a minimum.

Well designed Hi-Fi mains extensions (such as the ones I've developed) are designed to deal with much higher levels of current than you will ever get in a domestic Hi-Fi system. A few years ago I measured the power consumption of an older (and relatively power hungry) Quad system. It draws about 1 amp with a CD player. You could add a turntable and cassette deck and still draw less than 1.3 amps with eight items of Hi-Fi. If you had TEN such systems it still wouldn't reach the 13A rating of the block or your ring main.

Quad II Power amps x 2	859.00
Quad 22 preamp	24.70
Quad FM tuner	37.30
Quad AM II tuner	24.00
Tandberg TCD-310 cassette	82.56
Yamaha CD29 CD player	49.94
Garrard 301 turntable	113.59
	1.19

Power used by a Quad system

The 13A rating means that you can draw up to 13A safely because there will be only a slight heating of the wires and contacts. It doesn't mean that drawing 13A is unsafe in any way. There is a huge overload margin built in for safety. People often say that their equipment (usually a power amp) must be drawing a lot of current because it runs hot. This is just not the case. The equipment doesn't heat up instantly, it heats up to the designed running temperature over a period of time.

It makes good sense to connect all of the components in your Hi-Fi or Home Cinema into an extension: as well as being neat and convenient, using an extension block means that your system is effectively connected to your mains circuit at one point – known as 'star powering'. This helps your system to perform better than connecting your mains leads to several wall sockets at different points on the mains circuit.

Plug Fuses

UK mains cables are supplied with 13 amp fuses in them. The fuse is there to protect against mains cable damage; it will blow if you accidentally cut the cable, preventing electricity from flowing through the cable. A 13A fuse is right for this job, a 3A wouldn't be safer. Do not be tempted to replace the plug's 13A fuse with a 3A fuse - it will only degrade sound further than the 13A one has by increasing the impedance. We are, I believe, the only country in the world that puts fuses in mains plugs.

The 13A fuse cannot protect against faults in the Hi-Fi equipment because currents are far too low and the impedance in the mains cable is much too high. For equipment protection, rely on the fuse the manufacturer fitted into the equipment. This will be a fuse of about 180mA to 500mA for preamps, CD players, cassette decks, etc. and a 1.5A to 3.15A fuse for power amps. (I haven't used fuses in my Hi-Fi system's mains cables for 25 years, though of course we can't recommend you remove your fuses.)

Magnetic Circuit Breakers (MCBs)

Circuit breakers are the 'trip switches' that protect each individual circuit in your consumer unit (e.g. lights, cooker, downstairs sockets). I recommend circuit breakers in consumer units because they sound much better than fuses and are more effective as safety devices. They provide a better contact and therefore have a less negative impact on impedance. A domestic consumer unit will probably have three or more MCBs.

If you have an old type fuse box you may be able to upgrade the fuse that supplies the ring circuit that your Hi-Fi system is connected to by getting a circuit breaker of the same make as the box. Most manufacturers supply replacement circuit breakers and holders for their older fuse boxes. You must first switch off the supply at the main isolating switch and then remove the fuse and its holder to replace it with the circuit breaker and holder.

Connection Quality

All mains connections benefit from being clean and being kept clean. Physical and chemical changes in the surface conditions of electrical or electronic connectors are a primary cause of degraded performance of Hi-Fi systems. When contact surfaces are exposed to the atmosphere, non-metallic films are formed, inhibiting conductivity and increasing distortion and noise. Clean and protect all the contacts and switches in your system by using a good contact cleaner. I strongly recommend DeoxIT™ and DeoxIT™ Gold contact enhancers which are used by most manufacturers and which, in my experience, are by far the best. Below are some basic maintenance jobs that should be carried out about once a year. They cost very little and can make a significant improvement to sound quality.

Ensure that the socket screws securing the wiring in all the sockets in your Hi-Fi circuit are tight. **Beware -** before removing sockets, switch the mains off at the consumer unit.

If your consumer unit has circuit breakers, you can improve the sound quality by cleaning the breaker contacts. Just spray the contact cleaner into the circuit breaker – there's always a hole in it somewhere - and switch it on and off several times. Whilst the power is off at the main isolating switch, tighten up all screws securing all wires in and out of the consumer unit. A customer recently told me that this was the biggest improvement he'd made to his system, bar none!

Beware - Note that the wires on the INPUT side of the isolator switch are still LIVE (even when the unit is switched off), use an insulated screwdriver when tightening the screws.

Loose connections and unnecessary connections will also raise impedance and play havoc with your sound quality. Switched sockets, for example, sound worse than unswitched ones because the switch adds an unnecessary connection which increases impedance. All mains sockets sound different - often each socket in a double one sounds different! - and switched ones sound particularly bad. Replace switched Hi-Fi wall sockets with unswitched ones and check all the connections for tightness and security. After extensive tests (yes, I really do test everything!) I found that the MK logic sockets are the best sounding, and we can provide them ready treated with DeoxIT® for ultimate performance.

A customer has reported to me that he'd heard that unswitched 13A mains outlets were no longer legal. An electrician, a home surveyor and his local DIY superstore had all told him the same story. And a story is all it is. There is no truth in it at all. The Electrical Regulations have not changed and unswitched sockets are still approved, legal and available. In my opinion, switched sockets are less safe than unswitched ones because you are relying on the switch being left in the off position. It is easy to make a mistake so that it is on when you think it is off. It's much safer to have it unplugged.

Residual Current Devices (RCD)

The Residual Current Device (RCD, RCB, RCCB - same thing) is a kind of circuit breaker that protects by sensing for a fault condition in the balance of current between the Live and Neutral conductors. It is used to replace the isolator switch in a consumer unit. Most modern consumer units are now installed with RCDs as standard.

An RCD is a marvellous safety device for the use of portable outdoor appliances and kitchen appliances, etc. but (with one exception) is quite unnecessary for the safe operation of domestic Hi-Fi in a home. The exception is where the house mains supply is the old TT system that relies on a water pipe or earth spike for the safety earth fault path. Climatic conditions may raise the earth impedance (Zs) beyond the limit allowed in the IEE Wiring Regulations. TT systems only now exist in some country areas with overhead line supplies.

The problem with feeding the current to a Hi-Fi system through an RCD can be seen from Diagram 2.13, the current passes through a very small toroidal transformer. This is very undesirable! By increasing the impedance, the effect is to ruin good, clean, deep bass; make midrange sound thin and lacking in body; and add a hard, bright treble. It also affects soundstage size and dimensionality. That is why I have been advising against the fitting or use of RCDs (rather than an ordinary isolator switch) for the Hi-Fi ring circuit. I think they are usually an unnecessary evil when the circuit breakers in the consumer unit are very sensitive, very effective and very safe.

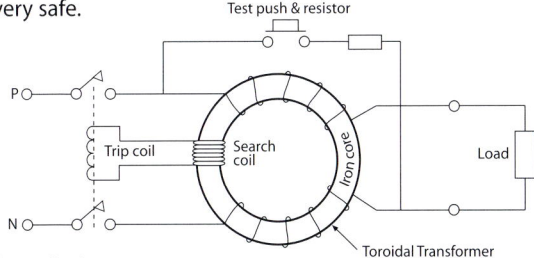

Diagram 2.13: Residual Current Device

A small number of enthusiasts have had difficulties with electricians who have insisted that RCDs must be fitted instead of the standard isolator switch when installing dedicated Hi-Fi consumer units and ring mains. The Regulations state, and I quote from the IEE 'On Site Guide', 16th ed., p 20 (carried by most competent electricians):

"30mA RCDs installed to provide protection to socket outlets likely to feed portable equipment outdoors should protect only those sockets."

Some electricians have interpreted the Regulations to mean that all downstairs sockets, wherever located, should be protected by an RCD. This is not part of the Regulations requirement. The operative words are *"likely to feed portable equipment outdoors"*. A dedicated Hi-Fi system socket in a sitting room with a Hi-Fi system permanently connected to it is not one 'likely to feed portable equipment outdoors' and so does not need to be protected by an RCD.

If you have an RCD fitted to your consumer unit and want to retain it and its safety factor for the rest of the house mains sockets but not the ring supplying the Hi-Fi system, simply ask an electrician to separate the Hi-Fi supplying ring from the rest and reconnect via a dedicated Hi-Fi consumer unit (see Diagram 2.16).

In the event that you cannot replace your RCD with an isolator switch, help is at hand thanks to Colin Lawson of Easom Electrical Ltd. in Maidstone. He told me that MEM make a type of RCD that senses the current balance in the toroid electronically and has a shorter, straighter run of thicker wire connecting the input to the output through the switch contacts. I bought one and tested it and it sounds a great deal better than the other kind.

The model is the A100 HE (100 amp version) RCD made by MEM and is available from any electrical wholesaler (though they will probably have to order it). If you are replacing your consumer unit or fitting a dedicated Hi-Fi consumer unit, you get a good deal by buying both the MEM consumer unit and A100 HE RCD together.

Mains Circuits

Your consumer unit will have a number of circuits serving different areas and functions in your home, e.g. one circuit for lighting, another for wall sockets. Putting in a dedicated circuit for your Hi-Fi or Home Cinema system is extremely beneficial. By increasing the isolation of your Hi-Fi or Home Cinema from your other circuits you greatly reduce the mains noise problem from household appliances (see **Mains Noise** page 35) and you ensure the best possible delivery of current to your system.

There are two basic types of circuit for house wiring: a radial (or spur) circuit and a ring circuit. A radial (spur) circuit is wired with a single cable running from the consumer unit and terminating at the furthest socket. A ring circuit consists of a continuous loop of cable running out from the consumer unit and back again with one or more sockets connected to it.

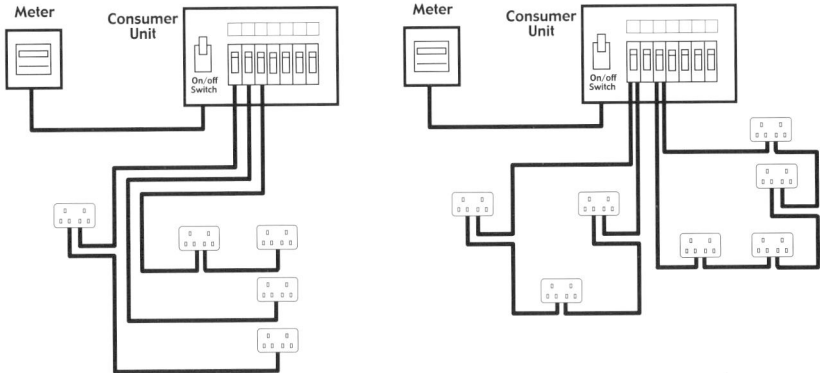

Diagram 2.14 Radial (spur) circuits   Diagram 2.15 Ring Circuits

The advantage of a ring circuit is that looping the wire back to the consumer unit doubles the current handling capacity of the cable by halving the impedance. The wire in a radial circuit must be much larger if it is to match the lower impedance of a ring circuit.

Although I have always advocated that dedicated Hi-Fi circuits should be installed as ring circuits, there are others who recommend installing spurs. The difference may at first sight seem academic, but in practice it is huge.

A number of enthusiasts have followed a major magazine's recommendation to fit multiple spurs, segregating analogue sources, digital sources, preamps and power amps. This scheme looks good (and 'technical') on paper but in practice the results are usually worse than the original ring main that it replaces. The results these customers have achieved including loss of bass and increased hum have been extremely disappointing and confusing to them. The experience of Mr Fletcher of North Yorkshire has prompted me to make my opinion on this clear and advise you not to repeat his experience. He spent a lot of time and trouble putting in the multiple spur system and then had to undo it all to put in a ring main.

Installing a dedicated ring main for your Hi-Fi

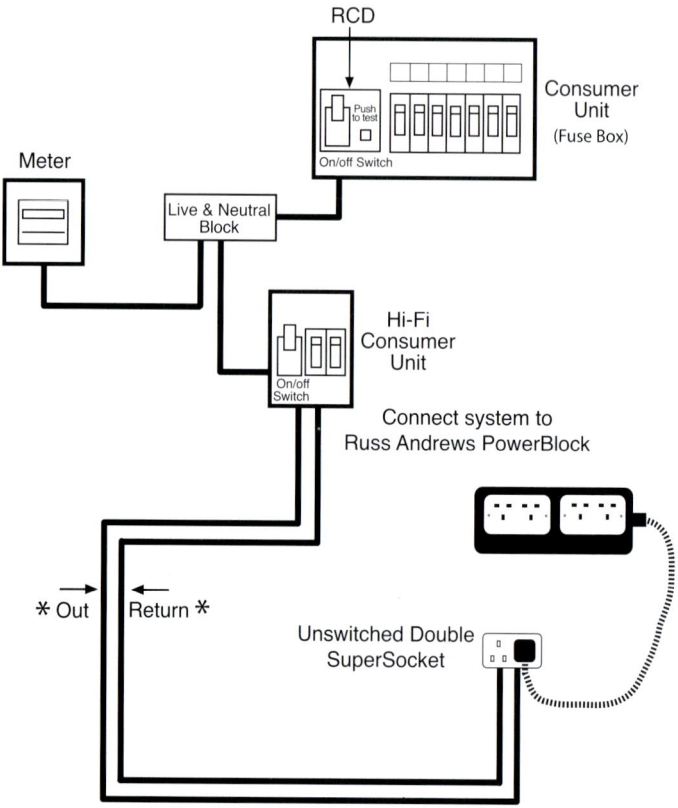

Diagram 2.16: Showing dedicated Hi-Fi local ring main

Note: The correct 'directionality' of these cables is that both run from the consumer unit to the destination socket - i.e. both direction arrows are at the end pointing to the socket.

See the Safety Notes on page 27.

1. Fit a dedicated Hi-Fi consumer unit wired in parallel to the existing consumer unit as shown. Do NOT add it to a fused outlet from the original unit. Some units sound better than others. We have tested many and found that those made by Hager, Proteus or MEM sound the best. You can get one from any electrical wholesaler.

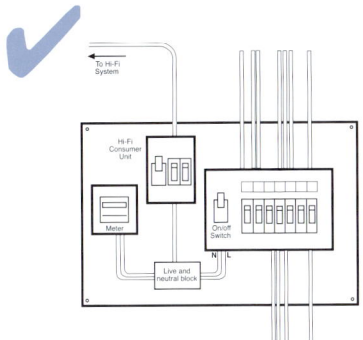

Diagram 2.17
Correct wiring:
Hi-Fi system consumer unit wired
in parallel to existing consumer unit

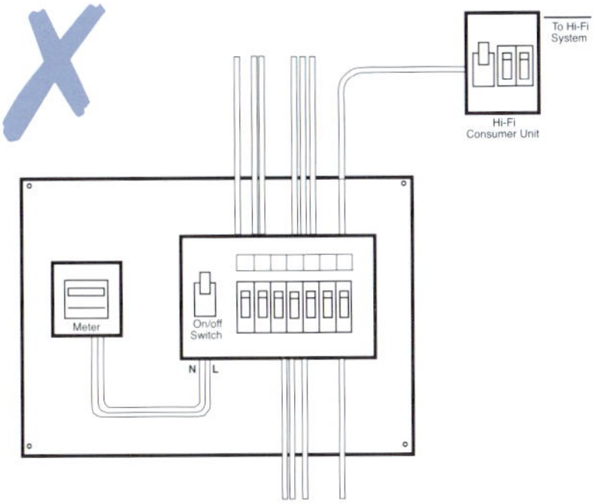

Diagram 2.18
Incorrect wiring:
Hi-Fi system consumer unit wired
to fused outlet from the existing consumer unit

2. Rewire with a low impedance ring main cable (such as Russ Andrews Ring Main) in a local ring circuit to an unswitched socket: see Diagram 1.16 on previous page. A dedicated Hi-Fi ring circuit should be wired with at least 4mm² twin and earth cable. (A radial or 'spur' circuit should have at least 6mm² twin and earth). The wire is normally broken at each socket and connected only by the screws in the socket.

3. First, of course, switch off power to the circuit and double check that the circuit is dead using either a mains tester or lamp. Remove each socket in turn and tighten the Live, Neutral and Earth screws. It is a good idea to test the circuit with a plug-in Socket Tester to confirm correct polarity, etc.

Chapter Three
Wired for Sound

Upgrade Path Steps 2 & 3: Interconnects & Speaker Cables

Obviously, cables must make the electrical connections between all the bits of hardware. It sounds simple enough, but doing it properly without performance penalties is actually very difficult. Let's be brutally honest – it's impossible.

This chapter explains how the cables that carry the signals in your Hi-Fi – interconnects and speaker cables – have an effect on the performance of your system.

The 5 Key Factors in Cable Performance

Everything about a cable affects the signal going through it. It subtracts from the signal (it is lossy and slow) and it adds to the signal (it creates distortion and is an aerial for RFI).

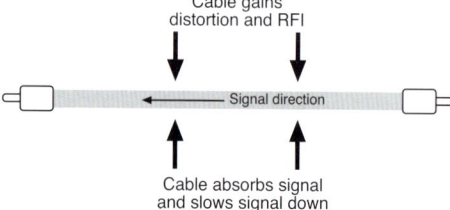

Diagram 3.1: How cables affect the signal

There are five key factors that determine how effective a cable is at transmitting a signal and rejecting interference: insulation, number of conductors, type and purity of the conductors, geometry and its connectors.

1. Insulation

Insulation (or the 'dielectric') is important for a number of reasons: it stops the cable shorting together, protects you from shocks from the bare conductors, helps to stop the conductors oxidising, and maintains conductor purity. The material used for the insulation (and even its colour) has a significant effect on the sound.

Most cable manufacturers use PVC insulation. Kimber Kable use a variety of insulation materials, but most commonly polyethylene and Teflon®. Both of these are better than PVC because they do less to the signal, but Teflon® is the superior material. Kimber have perfected a technique called 'Air-articulated Teflon®' where the Teflon® is applied to the surface of the conductor at high pressure. This ensures that no tiny air pockets are trapped between the Teflon® and the surface of the conductor, improving performance.

2. Number of Conductors

The number of conductors influences the sound in two ways. Firstly, by increasing the number of conductors, the resistance of the cable is reduced, making it better as a conductor. Secondly, in the case of Kimber's woven cables, more conductors mean that the cable weave is enhanced and is more effective at cancelling interference. The more effective the cancellation of interference, the lower the distortion levels and the more musical the sound.

3. Type and Purity of Conductors

Silver is generally a better conductor than copper, but be careful when using this as a deciding factor because it does depend on the purity. Whilst silver wire for interconnects and speaker cables should sound better than copper, this will only be the case if the silver is ultra pure - at least 99.99% pure! Ordinary 'sterling' silver is only 92.5% pure and sounds worse than good copper. This is why many American reviewers advise against silver cables, characterising them as sounding hard and brittle. I can confirm that these low purity silver cables sound inferior to the best copper cables. Silver can, however, be worked to a better finish than copper, which is why some of the very best mains, interconnect and speaker cables use 99.99+% pure silver cable in their construction.

Although silver plating sounds like a very attractive cheaper option (because signals travel only on the skin of a conductor), and is used extensively by some manufacturers, it isn't very stable because the interface between the plating and the base metal degrades comparatively quickly. A silver plated cable that sounds good in a magazine review may sound awful three or four months later!

Kimber Kable are the only cable manufacturers, to my knowledge, that oversee the production process of its conductors all the way from smelting stage to finished product. This enables them to ensure its purity, and being based in Utah, USA - one of the driest places on earth - helps ensure this purity is maintained!

4. Geometry

One of the most important features of a cable in a Hi-Fi or Home Cinema system is its geometry. Cables can comprise just one solid conductor or several strands. They can run straight, twisted, ribbon or shielded. The way a cable is constructed determines whether or not it simply acts as an aerial for radio signals (causing interference) and so affects its performance considerably.

Kimber Kable use a unique patented woven technique for each of their stranded cables to cancel Radio Frequency Interference (RFI) in a highly effective way (also see **The Solution**, page 23) . The weave ensures that the cable presents no coherent antennae pattern, resisting the pick-up of airborne RFI.

Furthermore the weave, coupled with the super-conductivity of the conductors, shorts out RFI that is already in the system. This design is so effective because it does nothing to degrade the signal travelling along the cable, unlike shielding and RF chokes which can interfere with the cable's natural magnetic field and degrade the signal.

High frequency signals travel along the skin of the cable so Kimber use several strands of wire in each insulated cable to increase the surface area. Their VariStrand™ technology uses seven strands of different diameters to spread this skin effect evenly and to help cut down resonance built up internally in the cable (see Diagram 3.2 below).

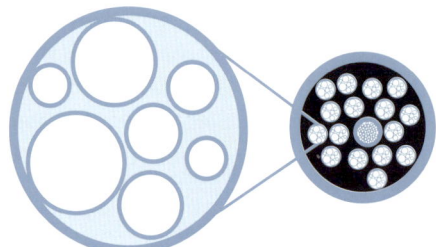

Diagram 3.2: Cross section of VariStrand™ cable

5. Connectors

It's no good designing a great cable if the connectors you use compromise its performance. A good quality connection must effectively couple the cable to the connector and the connector to the component it is plugged into.

I've found that gold plated connectors, as a general rule, are not always the best performing ones for two reasons. For starters, the gold plating can be so thin and impure that it wears off quickly. And then you have the problem that corrosion can build up between the plating and the base metal it is applied to, so it starts to sound harsh and distorted over time. Gold plated plugs that do perform well often feature a complicated multi-layer plating process to stop the oxidisation leaching through to the top layer and are, by necessity, expensive.

Cable Directionality

All cables - yes, **all** cables have signal directionality.

By that I mean that in one direction the sound is slightly louder, has lower distortion, is cleaner, smoother, sweeter, has deeper bass, and overall wider dynamic range. The amount of directionality, or the difference between one direction and another varies from cable to cable.

If a cable is labelled for directionality then you should connect it up so that the arrow points in the same direction that the signal is moving in, i.e. from your CD player to your preamp.

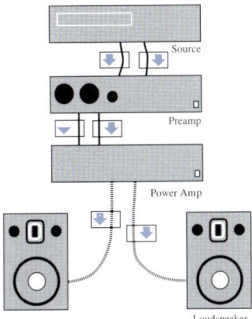

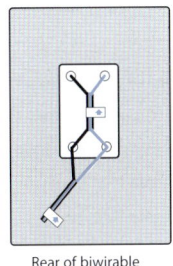

Rear of biwirable
loudspeaker

Diagram 3.3: Cable directionality - the arrows follow the direction of the electrical signal.

Diagram 3.4: Jumper cable directionality - the arrows follow the direction of the electrical signal.

If a cable isn't labelled as to its directionality it is very easy to directionalise with a listening test. Choose a piece of vocal music that has a bright nasty edginess to the voice. Listen to a short passage (5 to 10 seconds is all you need), reverse the cable under test and repeat the passage. You will be instantly aware that in one direction the sound is brighter, nastier, edgier, colder, etc. It is easier and less confusing to do one channel at a time. Don't do too many repeats if you aren't sure because it just becomes too confusing, try a different piece of music instead.

As you work through the directionality of all your cables you will find other changes that only become apparent in stereo. The sound stage will become more stable, phase accuracy improves, three-dimensionality becomes more pronounced and the whole system will sound more relaxed and musical.

Yes, before you ask, all the bits of wire inside the equipment are directional and it would be wonderful to be able to get them all right. Those of you who are good at Hi-Fi DIY can easily tackle that task, but don't worry if you can't. Every cable that you get right is important and is a positive step forward in the sound of your system. It is a free upgrade and so worth the effort.

Your work on improving your mains quality will make directionality more obvious, so it may well be the reason why you may have noticed an increase in edginess. The cause of the edginess is not in the mains upgrades but in the signal cables (cables carrying the signal, i.e. interconnects and speaker cables); the better mains quality is simply making the symptoms obvious. Don't blame the messenger for the message!

Burn In

When you plug a new cable into your system, don't expect it to perform at its best immediately. Whilst you will hear some immediate improvements, particularly with mains cables, new cables tend to sound bright and lacking in bass (especially if they are solid silver cables). They need time to settle down; I call this time 'burn-in'. The burn-in process takes up to 500 hours. That's a long time if you only switch your system on for a few hours each night. Burn-in is best achieved by having the system switched on 24 hours a day, then the new cable will be properly burned-in after about three weeks. Leave some music playing, for example, radio from your tuner, to speed up the burn-in process (you can turn the volume right down unless you are burning-in speaker cables).

Don't expect the new cable to simply improve gradually getting a little better each day. It isn't like that at all! The process goes through cycles where it may sound better one day than it does the next. Different areas change independently too; the treble sweetens up fastest, the midrange may go harder for a while than it was at the start and the bass takes longest to extend, deepen and settle. Aspects like soundstage, atmosphere, information, musicality, etc. won't finally settle until the end of the process.

The message is obvious, don't judge a cable as unsuitable for you until it is fully burned-in and settled.

Optimum Cable Length

In answer to the oft-asked question, *"What's the optimum length for your speaker cable?"* Ray Kimber quips, *"In my experience, it sounds a lot better if it reaches the sockets, and the longer it is the more money I make!"*

But seriously, the answer depends upon the type of cable. Signal carrying cables (i.e. speaker cables and interconnects) are best kept to the minimum length required to reduce the loss of information. This is particularly important in the case of the cable from passive preamps to power amps where the maximum length you should use is one metre. See also page 66 **Analogue Interconnects for Amplifiers.**

Interconnects

The term 'interconnect' refers to the low level analogue and digital connections between all the electronic equipment in your system. You should upgrade your interconnects before your tackle your speaker cables.

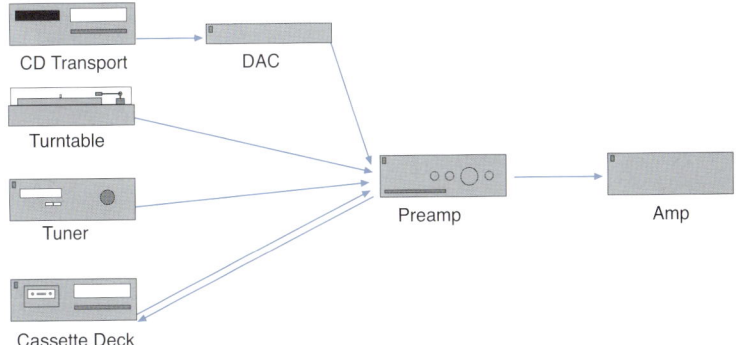

Diagram 3.5: Interconnect connections: the arrows indicate signal flow (or direction) through the system.

Analogue interconnects link source equipment such as a CD player or turntable to your amplifier; supplied in pairs (one per channel), they are also used to link your preamp to your power amp if you use this type of set-up. Digital interconnects are only used if you have a two-box CD player (or you are connecting a DVD player to a Home Cinema amp) to transmit a signal which remains in the digital domain to a separate digital to analogue converter. Analogue and digital signals have very different characteristics and a properly designed digital interconnect will not work as an analogue interconnect and vice versa.

When upgrading your interconnects, it is logical to first ensure you are getting the best out of your music sources, concentrating initially on those you use most frequently and want to upgrade most urgently. CD players usually fall into this category because they are harder to get enjoyable music from than FM tuners and turntables. I don't believe that they are difficult because the medium itself is particularly flawed. The major problem, I have found, lies not with the CD medium itself (though I welcome any improvements in specification and standards) but with the CD hardware. It is the hardware which is the main cause of justified criticism of 'CD sound'. CD (and DVD and SACD) players are all, regardless of cost, made down to a price not up to a performance. Commercial realities dictate this fact, not Hi-Fi aspirations.

I am often asked which quality of interconnect would be suitable for a given quality of system. There is a simple answer to this question: buy the best that you can afford.

One customer a couple of years ago tried one of Kimber's silver KCAG attenuated interconnect in a system costing less than a thousand pounds and was delighted with the results. He said the fact that the KCAG cost more than the CD player was irrelevant because the improvement brought was so big and so much better value for money than a similar cost upgrade to the player.

Analogue Interconnects for CD and DVD players

I have known for some time of the major problem of the very high CD player output voltage overloading the preamp/integrated amp input. It's this that is the real cause of the hard two-dimensionality that characterises CD sound. I launched a CD interconnect back in 1996 which 'attenuated' (reduced) the output signal as a simple and effective solution to this problem. However, it has become increasingly obvious that a number of other sources display this same overload problem: SACD, DVD, MiniDisc, DAT, FM radio tuners and digital radio.

Amazingly the overload problem still occurs even when the source amp are made by the same manufacturer! Not only does the overload add a hard two-dimensional sound, but it also makes the volume control very sensitive to minor adjustments - a minor point perhaps, but a really annoying one!

The standard CD/DVD output voltage is 2 volts and the correct input sensitivity of the amplifier (to allow for full dynamic range, differences in recording level, etc.) is 500mV (half a volt). This sensitivity works very well with no input overload, and a good range of control on the volume control. In practical terms this means that for normal listening levels on an average CD the volume control should be in the centre region of its range (12 o'clock on a clock face). This allows you to play your music very quietly when you want (for late night listening) and very loud at other times.

Diagram 3.6: Showing the normal listening level of a volume control

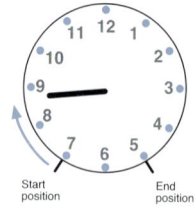

Diagram 3.7: Showing most people's listening level of a volume control

Most people now find that they can only use the volume control up to 8 or 9 on the clock face scale (or the first 10% of its range) because most amps are now designed with a 150mV input sensitivity. The manufacturers of course claim that their products do not overload at these sensitivities but the evidence proves otherwise. Most manufacturers, I think, do sine wave measurements to show that their 150mV inputs do not overload, but a digital source is not well simulated with a steady sine wave. Music isn't like that, it is a very rapidly changing transient 'burst like' signal. With this kind of signal most of these highly sensitive inputs exhibit what I call 'dynamic overload' because they can't follow the speed of transient accurately at CD output levels.

The effect of dynamic overload is to reduce dynamic range; mask differences in the music, cables, mains, support, etc.; flatten the soundstage from three to two dimensions; add bright hardness (or sometimes dullness); and reduce bass extension. Worst of all, it takes the musicality out of the music.

When a preamp or integrated amp circuit overloads, even mildly, the sound loses dynamic range and takes on an urgent, over-dynamic, 'grab you by the throat' speeded up character. This, unfortunately, is the sound that many people are used to hearing from a Hi-Fi system. It is understandable, therefore, that they think that this speeded up, 'dynamic' sound is 'right'. In truth, this sound is the product of overload distortion and nothing to do with the sound that was recorded on the disc. When you take away the overload distortion, their reaction to the improved sound is that it is 'slow' and lacks dynamics, and they think it is 'wrong'. I confess that I usually have great difficulty explaining that the new sound is a step in the 'right' direction. They may, of course, be right that the music sounds too soft and flat, but this is usually because other parts of the system have been selected because they mask the overload problem.

The volume control is just like a water tap, the number of turns varies the amount of water you get. The pressure of water in the water main controls how many turns of the tap are required. (The water pressure is equivalent to voltage in electronics.) We have all probably experienced the effects of too high a water pressure in a wash hand basin. You turn on the tap a little to wash your hands and the pressure is so great that the water shoots straight out of the basin and down the front of your trousers! To cure it, all you need do is fit a pressure control valve (an attenuator) on the pipe to the tap and turn the pressure down. You can easily adjust the pressure valve so that you can get a controllable dribble to fill a glass or gush to fill the basin depending on how many times you turn the tap.

On a preamp or integrated amp the volume control just acts as a tap for the volume you get out of the speakers. However, in most amps it is being used as the 'pressure valve' as well, and the result is a similar lack of controllability found in the water tap under too much pressure. The amp manufacturers are trying to get the volume control to do two different things at once; to act as an input source level attenuator (pressure valve) and as a control for setting the listening level (volume). The result is that it does both jobs very badly. In professional equipment the functions are kept separate and called 'gain' and 'level' respectively. Guitarists use the 'gain' control to deliberately cause overload effects whilst keeping the 'level' down to the volume they require. In a Hi-Fi amplifier the last thing you want is overload, so you must reduce the voltage (pressure) to the volume control (tap).

To cure the problem, you need to reduce the output from your CD player (or if you use a two box CD and DAC, from the DAC) to match the input sensitivity of your amplifier. If you are lucky enough to be able to adjust the output level on your CD player, reduce it so that you can turn the volume control on your amplifier about half way for average listening. If you don't have a volume control on your CD player, the best bet is to fit an analogue interconnect with attenuation, or use your existing interconnect in conjunction with attenuation adaptors. If you opt for one of the Kimber interconnects with built in attenuation, as well as reducing the CD overload problem, you also get the benefits of Kimber's unique cable weave to reduce Radio Frequency Interference.

Once the overload problem is cured, you will hear increased dynamic range, increased dimensionality, smoother sweeter treble, deeper bass and, above all, much more musicality. You will also find that you can now use much more of the range of the volume control. The range of control is more necessary because the increased dynamic range available will mean that you need to use the control more to be able to set the right level for the recording you are listening to.

To work out what level of attenuation you need, find out what the input sensitivity of your amplifier is (often printed in the specs as an mV figure.) Then find out what the output level of your CD player is - usually 2.0 volts.

Now consulting the Attenuation Guide read across to find out what attenuation you need. So, for example, a CD player with a 2V output and an amplifier with 300mV input would need an attenuation level of -11dB.

Attenuation Guide

Source equipment output \ Sensitivity of equipment input	150mV	300mV	500mV
2.0V RMS (5.64 v PP)	-16dB	-11dB	None
2.4V RMS (6.78 v PP)	-16dB	-11dB	None
2.9V RMS (8.2 v PP)	-19dB	-14dB	None
4.0V RMS (11.3 v PP)	-22dB	-19dB	-11dB

Which equipment needs attenuation?

CD and SACD players have outputs of 2V or more; so attenuating their outputs is a must to hear these sources at their best. This also applies to two box CD players with a separate Transport and DAC; separate DACs have some of the highest output levels we have come across, typically anywhere between 2.4V and 4V. HDCD compatible players, for instance, have a 4V output on HDCD.

As with CD players, the analogue output of DVD players is 2V or higher, and when watching DVDs or playing CDs (with the sound through the analogue input on your system) they will benefit from using an attenuated interconnect. No attenuation is needed on the picture output or the digital 5.1 sound output when routed to your surround sound receiver or processor.

Tuner output levels have gradually risen from the 300mV of the old analogue Quad FM4 era, through 500mV and 750mV of the mid 1980s, 1V in the early to mid 1990s, to the current DAB digital tuners with a 2V or higher output to equal CD player outputs.

With preamp input sensitivities of 150mV the older tuners will match well but the tuners with 1V to 2V outputs will overload just like the CD players. The overload will not be as bad because most broadcasts are now compressed and limited, but you will have the problem of lacking control to adjust the volume levels to suit low level listening needs. Attenuating the outputs will have an obvious value in this situation just as they have for CD players.

As with tuners, the output levels of cassette decks have changed over the years and the most recent models have the same high output levels as CD players and will benefit from an attenuation on the playback lead. MiniDisc and CD recorders all have 2V outputs or higher and should be treated just the same as CD players. In all these cases, the Record lead will not need attenuation.

Russ's Tip

Most people are completely unaware of the damping/degrading effect of connecting the Record circuits on a recorder to a preamp or integrated amp. The Record circuits load down the signal at the input circuit of the preamp and reduce the 'life', information, 3-dimensionality and musicality of the signal.

If your preamp does not have a 'Tape Defeat' switch to prevent this loading effect except when recordings are being made, then you can simply unplug the Record leads from the preamp. When you want to record, just plug them back in.

Analogue Interconnects for other sources

The vast majority of tone arms (pick-up arms) on turntables are fitted with general purpose coax arm cables. They are made 'general purpose' so that they will work satisfactorily with all cartridges whether they are moving coil or moving magnet types. Moving magnet types work into high impedance inputs (47K at least) and need coax or screened interconnect cables to prevent hum. Moving coil cartridges, on the other hand, are low impedance sources working into low impedance inputs (30 ohms to 470 ohms) and can deliver improved performance with unscreened cables.

Analogue Interconnects for amplifiers

I am often asked which is better - balanced or single-ended connections between pre and power amps. The answer is that it rather depends on how the equipment has been designed. Some manufacturers, like Audio Research, Krell, Mark Levinson, design their circuits as balanced circuits throughout. If this is the case, then they should be connected together with balanced cables because the unbalanced (single ended phono) option will be a compromised lower performance connection. Some other manufacturers have unbalanced circuits but satisfy market demand to offer 'balanced' connections. These are often fudged pseudo-balanced outputs that will not sound as good as the unbalanced phono to phono connections. Do not try to connect balanced to unbalanced circuits if you want the best sound.

Certain preamp/power amp combinations do require a screened cable to prevent hum. Some Quad power amps and valve amps, for example, have a high input impedance that necessitates the use of a screened cable.

If you use a passive preamp, short leads from passive preamps to power amps are absolutely essential. One metre is the maximum length that should be used, otherwise the cable forms a high pass filter with the input impedance of the amplifier: a filter that is variable depending on the setting of the volume control of the passive.

If you have an active preamp then you may be able to run long interconnects without any performance penalty. I say 'may be' because preamps vary in their ability to drive long lengths of cable. You need a preamp with low output impedance (75ohm to 100ohms) that has a healthy output current. Most output stages, unfortunately, lack the current drive necessary. The manufacturer may be able to tell you if his preamp is really capable in this requirement.

If your preamp is good at driving long lengths of interconnect, then it may well pay you to site your amp or amps close to the speakers so that you need run only short lengths of speaker cable. This will enable you to benefit from using a much higher quality cable as the cost of a short length will be more affordable.

If you want to Bi-amp or Tri-amp your speakers then the long interconnect/short speaker cable option is very desirable. This is the case even if the speakers are only a few metres from the main equipment rack.

Analogue Interconnects for Home Cinema processors and amplifiers

Once decoded, a Home Cinema Processor outputs its surround sound signal at line level to external amplification. This means you'll need at least 5 mono runs of analogue interconnects to get basic surround sound.

Digital Interconnects

The digital interface between a separate CD transport and its partnering DAC - or between your DVD player and Home Cinema amplifier - is a very critical area.

Added to the usual problems of conductor type and quality are the constructional problems of accurate impedance over a wide bandwidth and the need for a very wide frequency response. Digital interconnects can be either optical or electrical. It is worth remembering that optical digital cables can give better performance than electrical digital cables in low and mid priced equipment and runs over 5 metres. The performance variations between makes of fibre-optic cable can be very large. A cheap one is not a good indicator of how good a good one might be in your system.

For best performance of SACD and CDs use the analogue output from the player into your Home Cinema Amplifier or Hi-Fi system and only use the digital output for DVDs where the 5.1 surround sound information is available.

Use a digital interconnect to link a digital recorder directly to the output of the source. The quality of interconnect needed depends, of course, on the importance of the source.

Aerial Interconnects for FM and DAB tuners

A coaxial cable comes from your aerial or wall socket and carries TV and tuner signals. It carries picture and sound as a 'radio frequency' signal, which is why it's often referred to as an RF cable.

It is well worth the effort and cost of replacing the whole down lead with the best aerial cable for the improvement in stereo radio performance.

FM tuners are very dependent on the aerial signal they are fed with. Even if there is sufficient signal strength to achieve full quieting (lowest noise level), the design and siting of the aerial and the quality of down lead will have an enormous effect on distortion, dynamic range, sound stage and low level information.

Don't skimp on the aerial, the benefits of a good one are well worth the investment. The BBC still broadcasts exceptionally high quality live concerts on Radio 3; drama on Radio 4 and many Radio 1 and 2 programmes are of high sound quality, as are Classic FM and Jazz FM. Local radio varies a great deal but many live studio broadcasts are very natural and realistic.

Much has been said of the quality of digital radio (DAB) and I admit that you can still get better results using FM for music! The choice of stations is very wide with DAB, however, so it's worth doing all you can with your DAB tuner to get the best results. As with FM, your aerial leads are very important so replace the aerial cabling that connects to your aerial with higher quality coax cable. Many DAB tuners – even the cheaper ones – have a digital output (optical or coax) and you often get better results if you connect them via this connection to either a DAC or the digital input on a Home Cinema amp. And you can often get even better results from DAB radio if you listen through your TV Freeview box or Satellite – the bit rate is higher for many stations through your TV's tuner box.

Aerial interconnects for your Home Cinema

Whilst this book is primarily about Hi-Fi systems, it's worth mentioning that television and video pictures are as dependent on the aerial signal they use as Hi-Fi radios. There are one or two very low loss aerial cables available now that outperform the standard and 'satellite' cables normally fitted to TV, FM and satellite aerials.

Even the short length linking your video to TV will bring improvements in picture and sound quality, with sharper colours and a more realistic sound. The more lengths of the standard cables you replace with better cable, the greater the improvements.

Interconnects for Home Cinema

Home Cinema interconnects can be split into those which carry picture signals, those which carry audio signals and those, such as SCART, that can carry both sound and picture together.

Now that HDMI and DVI connections are common on DVD players, processors and flatscreens and projectors, big improvements can be achieved with the best HDMI and DVI cables. Further improvements in picture quality can be achieved by using the Meridian HDMAX 121 signal extender. It is very effective, even on short runs of cable.

Speaker Cable

If you want to hear the full potential of your system then a high quality speaker cable is critical. It is important that you prevent information being lost or distorted here as you're not going to be able to recover it later. You're looking for a cable that is capable of handling high voltages and currents (higher than those carried by the interconnect cables), and that minimises any deterioration of the information being carried by it.

Speaker cables are usually the first thing that comes to mind when the urge to upgrade strikes. But beware! Important as speaker cables are, they should really come after you have upgraded the cables earlier in the chain (and, yes, that means starting with the mains cables!). Just think about the fact that they can't improve on what comes before, they can only prevent further degradation.

If you go straight to speaker cables without solving problems earlier in the chain you're only going to be able to hear a fraction of the improvement that the cable can offer, and that's really bad value for money. Worse than that, you may in fact end up choosing a cable that hides the problems and so sounds superficially better, when you really need a cable that reveals problems (See page 17: **Learn to Distinguish between a 'Symptom' and a 'Root Cause'**). After all, you can't fix a problem if you can't identify it!

There is one situation where you might be wise to consider changing your speaker cable ahead of other cables in your system - if you are still using bell-wire. This is the exception that proves the rule. In this circumstance, any of the Kimber speaker cables will offer a huge improvement, and you can always move up to a better quality one later. Just bear in mind the 'hiding symptoms' problem and don't expect miracles.

Noise through the back door

Noise picked up by speaker cable degrades audio system performance in a number of ways. Since it is apparent that noise picked up by speaker cables would not be of sufficient power to directly drive the speakers, the recognition of this noise pickup as a problem has not been realised in the past. Noise picked up on a speaker line can actually be amplified by the gain of the entire audio system. (For more details on **RFI and amplifiers** see the Appendix I at the end). As previously discussed woven cables are very effective at combating noise (see discussion of RFI on page 22).

If you are one of those people who insist on choosing cables by technical specifications Ray Kimber has a piece of advice: *"A superior speaker cable is one that allows some increase in shunt capacitance in order to reduce series inductance and resistance."* Cables with low resistance perform well only if they also have low inductance.

Beware of judging a cable on one parameter only. The only exception to this is the most important parameter of all - the sound.

Choosing the Correct Speaker Cable

As with interconnects, I am often asked which quality speaker cable is suitable for a given quality of system. The answer is the same - buy the best you can afford! I usually explain that 4PR, the least expensive Kimber speaker cable, won't disgrace a High-End system because its character is inherently musical. At the other end of the scale, we have customers with more modest systems who use 8TC, Monocle™ or Select speaker cables because these cables get so much more out of their systems.

About Bi-wiring

Bi-wiring is very fashionable these days, promoted by Hi-Fi magazines and slavishly catered for by speaker manufacturers who supply their loudspeakers with terminals to allow bi-wire or even tri-wire connection. Don't get me wrong, bi-wiring certainly can be better, but only if you use a cable properly designed for it. The only real benefit in bi-wiring (or tri-wiring) is to double the amount of cable used, thereby halving the impedance and improving bass performance.

The problem is that many amplifiers find the load more difficult and so the sound is degraded in some way. People often experience the effect that bi-wiring improved the sound in some areas but degraded it in others (principally soundstaging and coherence). A properly designed bi-wire cable will avoid these problems, but will, of necessity, be more expensive. My best advice though, is to move up to a better single wire cable rather than to go for an extra pair of the same quality cables to get bi-wire operation.

Earthing your Speakers

As I mentioned in Chapter Two, some loudspeaker manufacturers are starting to introduce loudspeakers with an earth terminal. This is connected internally to the chassis of the drive units, and allows you to connect these sockets with green/yellow earth wire to the earth socket such as those fitted to high quality mains extensions or the earth pin in a mains plug to benefit from improved clarity and enhanced spaciousness.

If you don't have earthed speakers, you can test this upgrade fairly easily for yourself without having to modify the drive unit. When you remove the speaker grille, you will find the tweeter(s) and one or more drive units held in with screws. Tackle only the drivers and tweeters with metal bodies. Drivers and tweeters that have a plastic chassis do not need earthing!

Remove one screw from each driver and clean the metal around the screw hole to remove any paint. Strip the end of a piece of earth wire that is long enough to reach your mains socket (you are going to use its earth), make a small loop in the stripped end and refit the driver screw through it so that it makes a good tight connection with the chassis. If there is more than one driver on each speaker, you can earth link them together or run two lengths of wire.

Take the opportunity to check the tightness of all the driver screws but don't be tempted to over tighten them. Do both speakers and fit 13 amp plugs to the earth wire (using only the big earth pin!) so that you can easily A/B the difference.

If you like what you hear, you can make the connection at the back of the driver more permanent and hide the earth wire by making the earth connection at the back of the driver and running the earth wire to an extra socket on the back plate next to the usual speaker terminals. If you find the tangle of earth cable unwieldy, a Star Grounding Block (page 43) is perfect to link all your earth cables together.

If you do choose to use single wiring then you can get a very cost-effective upgrade by replacing the cheap and nasty jumper links that link between the terminals on your speakers to allow single wire use. My tests have shown that by simply replacing them with a high quality cable you can achieve clearly noticeable improvements in sound quality.

Speaker Cables for your Home Cinema

In a Home Cinema System the speaker cables are more important than they are in a Hi-Fi System because there are more of them and they are longer!

Speakers in a Home Cinema system set up can present a particular challenge. Long runs are often required to the rear speakers, and long runs can mean a reduction in quality. However, with a woven cable design this is not a problem. Extremely low resistance, coupled with RFI cancelling, make it an excellent choice for long runs where the sound reproduction is critical.

Chapter Four
Firm Foundations

STEP 4

Upgrade Path Step 4: Supports

The conventional wisdom on Hi-Fi supports, racks and stands is that all you need is lots of 'mass' or weight. Metal and glass are currently very fashionable but enthusiasts who really 'go for it' choose slate, granite or concrete! Indeed, Jonathan Scull writing in 'Stereophile' June 1999 says *"Remember, you're looking for high rigidity and whatever mass you can afford to purchase".*

Sorry Jonathan, wrong on both counts! (Unless you are looking for 'grab you by the throat', 'smack you in the face', punchy exciting Hi-Fi sound effects, in which case you should put this book down now and go for the metal and glass, high mass racks just like the magazines recommend!)

If you are still with me I assume that you, like me, are looking for the real, natural sound of music. And to get fast, clean, natural, neutral, wide dynamic range, three-dimensional music, you should really be looking for high stiffness and low mass. Why? The following pages will explain the logic behind my unconventional solution to equipment supports and acoustic feedback control.

Why do supports matter?

Why should it matter what you use to house or support the equipment in your Hi-Fi system? The reason is 'acoustic feedback' from loudspeakers which seriously degrades the sound quality a Hi-Fi system makes - regardless of how good or expensive that system is. The usual choice of high mass (usually metal and glass) stands, racks and cabinets makes the problem worse not better. To find out how and why, read on!

It's only electronics, how on earth can it tell the difference? Surely any possible difference would be too small to notice? Good questions and ones you can best answer for yourself by conducting a simple experiment: try the Cushion Test! (page 75)

You should hear a big difference between 'cushion in' and 'cushion out'. (If you didn't then this is your second opportunity to close this book and give it to a friend!) Don't worry at this stage whether the sound is 'better' or 'worse', we're just proving that there is a difference.

Try this experiment with your preamp, power amp or integrated amp, speaker, etc. You will find that every part of your system changes its performance depending on what it is standing on.

When I first heard the difference that a support made, two questions immediately demanded answers: why did the support make a difference? And consequently: what was the best support possible?

I have to confess that I made two serious blunders at that point. I didn't answer the first question properly and went straight for the conventional solution - high mass (slate in fact). I was using a turntable to test different materials. What I didn't realise was that I had a faulty pickup arm on the turntable. The slate sounded better, but only because it hid the problem. I fell for it hook, line and sinker! This experience taught me two important lessons:

1. Never, ever launch a product without fully researching and developing it.

2. Never, ever confuse symptoms with causes! (See page 16, **5 Keys to Successful Upgrading** for more on this.)

All my early work was with turntables: the source component most sensitive to the support problems and therefore the most revealing of the differences. It made the development process quicker and easier being able to hear the differences in construction and materials so clearly.

It also allowed me to prove that I had a universal solution to the problem of supports rather than one limited to a particular product or situation. I found that it didn't matter whether the turntable was cheap or expensive, heavy or light, belt drive or direct drive or any combination - the effect of the correct solution (my Torlyte® turntable stand) was totally beneficial and the same across the board. Subsequent work with CD players, preamps, power supplies, amps etc proved the point and confirmed the 'rightness' of the solution.

Any discussion of the causes, effects and solution of the Hi-Fi system support problem must begin with a clear analysis of the mechanics involved. This is not rocket science but the simple mechanics (physics) everyone was taught at school. The effects those simple mechanics have are very interesting and lead to problems in the electrical or electronic domain.

The Cushion Test

1. Slip a cushion between your CD player and the surface it stands on. Play a few seconds of music.

2. Take the cushion away and play the same music again.

The Turntable Test

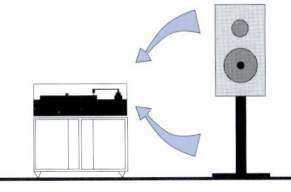

1. Place a record on to the stationary platter and put the pickup on the record as if you were going to play it. (Unplug your turntable from the mains supply if the platter starts revolving when you move the pickup onto the record.)

Diagram 4.1: Feedback from the speakers

2. Turn the volume up to full (yes, full volume!) and listen to the sound from the loudspeakers. There will be some hum and noise but usually nothing more.

3. If you now tap the turntable plinth with your knuckle to simulate the effect of feedback, you will hear a 'thump' sound from the loudspeaker. The character and quality of that 'thump' tells you how your turntable and also the whole system responds to acoustic feedback.

In most cases you will hear a long, slow, sluggish boom noise from each tap. In bad cases, the tap can start 'howl round' feedback that can only be stopped by turning the volume down. Ideally, all you should hear from the speakers is a clear, clean, short, sharp tap coloured only by the box sound of the plinth.

4. Tap different parts of the turntable, plinth, stand, etc. The contribution of various parts can be very interesting and instructive.

When you have recovered from the shock of this little experiment, consider the effect that the sound from the speakers is having on the rest of the system.

Acoustic and Electrical Feedback

I went into some detail in Chapter One about acoustic and electrical feedback, but it's worth repeating some of it here in the context of equipment supports. Remember that a Hi-Fi is not a simple chain of components with a source (e.g. turntable, CD player) at one end and a pair of loudspeakers at the other end. The two ends (and every point in between) are connected together acoustically by the sound from the loudspeakers passing through the air, the floor and, crucially, the system support stands - this is 'acoustic feedback'. Turn back to page 15 to see Diagram 1.1 showing this feedback.

As the acoustic feedback enters each piece of equipment it becomes 'electrical feedback'. How? Well the electronic components inside your Hi-Fi equipment (e.g. capacitors, diodes and resistors) are 'microphonic'. That means that they produce small electrical signals in response to mechanical vibrations such as acoustic feedback. These electrical signals (electrical feedback) are not in themselves very large, but in relation to the audio signals being processed they are very significant indeed. They mix with the audio signal and pollute it audibly.

And to make matters worse, the feedback effect is cumulative. The energy content and frequency balance of the feedback signal is added to the music signal passing through the system and modifies it cumulatively. This cumulative effect is a classic vicious circle where the system sounds worse and worse as you play it louder and louder.

This acoustic and electrical feedback link seriously compromises the performance of your system by creating 'time smear' distortion and 'colouration'. The result is an increase in tonal colouration, loss of information and detail, a reduction in dynamic range and, worst of all, a marked reduction in musicality caused by time smear. The turntable is the most sensitive to the progressive feedback effect, and a very revealing test (see page 75) can be performed if you have one in your system.

Colouration and Time Smear

The first thing you may notice is the effect feedback has on the tonal colour of the music (colouration). However, the worst and most destructive effect feedback has on the music is to cause 'time smear'.

The feedback signal is like an echo; it is a delayed repeat of an earlier sound.

Diagram 4.2: Long delay, as in church acoustics

Diagram 4.3: Short delay as in your Hi-Fi system

The confusing effect of this can be heard in its most extreme form in a large church or cathedral. In this situation an experienced speaker will pace his delivery to allow for the echo and so maintain intelligibility.

In a Hi-Fi system however, even the small amount of echo visible in Diagram 4.3 is a very big problem because it causes 'timing smear' that destroys musicality.

Timing is very important in this world; never more so than in music. After you learn to play the right notes in the right order - timing is everything. It conveys emotion and meaning. Timing is the difference between a good musician and a great one! Timing **IS** musicality. If timing is so important in the creation of music, then logically, timing must be equally important in a music reproduction system.

The time smearing that occurs in a Hi-Fi system 'bleaches out' the strength of the rhythm and destroys the timing subtlety of the playing. After taking steps to reduce timing smear in a system, listeners have made comments about the ability or talent of the musician: *"I didn't know he could play that well"* and *"I was bored by that piece of music before, but now I quite like it!"*

This is what is so important about time smear (or more technically 'Time Domain Coherence'): it controls enjoyment of the music itself rather than just affecting the sound of the music.

The message I want to get across is the extent to which equipment supports affect rhythm, timing, bass depth and colouration; and the fact that the conventional Hi-Fi recommendation to use metal, glass and high mass (weight) is completely wrong. Not just a little bit 'off message' but wholly, totally and completely wrong!

I'm sure you find it difficult to believe that so many people can be so wrong. And I agree that it is difficult to believe. The problem is that people have a tendency to accept things that they have read or been told without questioning it or researching it for themselves. Because of this, inaccuracies and errors get repeated. If these inaccurate assumptions are then used as the basis for further research it is easy to see how we can be sent off course. If you do think about this problem of acoustic feedback and apply some very elementary physics to it, the real answer is staring you in the face!

Dealing with Acoustic Feedback Energy

The acoustic feedback energy from the loudspeakers is the problem, so we must somehow 'deal' with that energy to stop it causing trouble. Unfortunately, we can't just destroy it because as Sir Isaac Newton so elegantly proved:

'Energy can be neither created or destroyed'

He also proved that:

'In every action there is an equal and opposite reaction'

We can't stop the energy getting into the system, but the energy itself isn't really the problem, it's the effect that energy has - time delay - that is the real problem.

The time delays occur because the feedback energy is slowed down and stored as it passes through and into each part of the system. It is the mass (weight) of each object that dictates how much energy is stored. The higher the mass of an object, the lower its 'resonant frequency'. (Every object has its own 'natural resonant frequency'; the frequency it vibrates at when you knock it.) The lower the frequency then the more energy is stored and the more the energy is slowed down. This slowing down means that the object will continue to vibrate (or resonate) for longer thereby adding to time smear.

Diagram 4.4: How high mass results in time delay

If high mass increases time delay, then low mass will decrease time delay. The lighter (lower mass) things are, the less energy they can store, the less they can ring, and the less they can delay the feedback energy.

Therefore, if we could make the system support materials very light we would have a system that stores little energy and delays it less. We will have raised the speed of reaction to the feedback energy and as a result reduced the time smear. We know we can't stop energy moving but if we ensure that it moves as fast as possible through the system, and that all parts of the system vibrate together and very fast, then the vibrating ceases to matter. To do this we must couple all the light weight parts of the system together in such a way that we allow the energy to pass freely from one to another, without coupling their masses together to create one big high (slow) mass.

Coupling the masses is a technique used by cinematographers filming microscopic single cell marine organisms. The camera has a motor that vibrates it so much that the organisms in the water tank would be invisible. The solution? They clamp the water tank to the camera so that it all vibrates together and so the vibration itself doesn't matter.

The Influence of Materials

There is, however, another problem caused by the feedback energy. The tonal character of the music (frequency response) is changed by the materials the feedback travels through. In other words, you can 'hear' the support materials in the feedback signal and it changes the sound of the music. You can hear the metallic ringing of metal stands, the 'dong' and ring of glass shelves, the thick heavy thump of medite (MDF), etc. Proponents of the conventional high mass support system go one step further, to make problems even worse. They say that the equipment should be 'isolated' from the shelves with soft 'squidgy' rubber feet. This has the unfortunate effect of trapping the feedback energy in the piece of equipment, thereby increasing the storage and delay!

A further and equally music destroying effect of all this weight and stored energy is to reduce the dynamic range of the music. It concentrates all the energy in a narrower band, delivering a sound that is louder, punchier and more up-front. Mid-bass is forward and dominant, and the treble grabs you by the throat!

On certain kinds of Rock music this sound effect is much favoured in Hi-Fi circles at the moment, but other kinds of music reveal it for the sound effect it really is. Unfortunately, it is the sound most people associate with a Hi-Fi system these days and that is a tragedy for enthusiasts and for the industry. Don't let this sound fool you - stand back and hear it for what it really is, the way ordinary non-enthusiasts hear it.

Your friends and neighbours should be stunned by the totally believable, natural sound your system makes, not laughing at you for wasting so much money on a loud obnoxious row! Believe me, though, once you change your approach and your system starts to deliver clean, clear, relaxed, natural music, the attitudes of family and friends will change dramatically. They will congratulate you on what you have done to the system, will enjoy listening to it, and will approve further upgrades. Your friends will visit more often and stay longer. I know these things happen because I have letters to prove it!

So, if high mass, metal, glass and rubber feet aren't the answer to equipment cabinets and supports - what is?

An Unconventional Solution

Sir Isaac Newton's laws of dynamics prove that we can't attempt to absorb the feedback energy without unacceptable penalty. We also know that we want to avoid energy storage and time delay; and to avoid distorting the feedback energy in any way (if the character of the feedback sounds the same as the sound from the loudspeaker then it is less obtrusive).

As I have demonstrated in 'Dealing with Acoustic Feedback Energy', page 77, the solution is to couple the system components to the floor rather than to decouple them. To do this we need some kind of stiff, low mass structure to support the equipment.

The low mass will mean that little energy is stored, and the stiffness is vital for raising the resonant frequency away from the problem subsonic region. Stiffness and low mass will together form a 'high pass filter' that naturally and mechanically reduces the bass content of the feedback energy.

All about 'Q'

What about materials? An important factor is the 'Q' of the material. A material's 'Q' is its willingness to resonate cleanly (and loudly!) at one frequency. Metals are good at that, they have a high 'Q', which is, of course, why bells are made from metal!

Where Hi-Fi is concerned, metal (particularly steel) is out as a suitable material for several reasons; although it can be used to make a very stiff structure its mass is much too high and so is its 'Q'. It concentrates the feedback energy into a narrow band at its resonance that then distorts the frequency character of the whole system. This compresses the dynamic range, increases time smear and has the effect of pushing the music forward. A metal stand, rack or table is very good at enhancing the reproduction of electric guitars, whilst unnaturally compressing the sound of the other instruments - exciting to listen to in the short term but quite artificial. For these reasons it gives non-Rock music a very strange, unnatural sound.

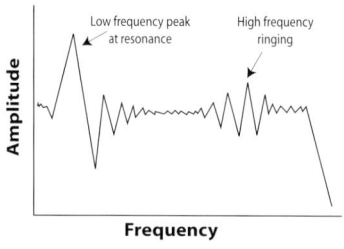

Diagram 4.5:
Typical frequency response of a material with a high 'Q'

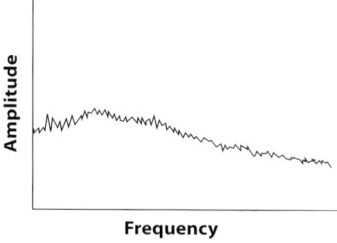

Diagram 4.6:
Typical frequency response of a material with a low 'Q'
(also known as a broadband response)

Steel has the additional problem of nasty high frequency harmonics that further distort the feedback signal. This ghastly scenario is made very much worse by the use of glass shelves. Glass is high mass (heavy) and has a low resonant frequency and high ringing harmonics. The result is one-note bass, forward lower midrange and grab-you-by-the-throat treble. Heard that sound somewhere? Yep - almost every Hi-Fi system you've heard!

Marble, slate and concrete slabs have their own unacceptable problems: huge mass and therefore huge energy storage plus ringing resonances. These materials are OUT!

Although it can be demonstrated that increasing the mass of a loudspeaker stand seems to work - increasing stability and damping the higher resonances of the metal - it actually stores energy, giving a resonant 'hump' in the low frequency response of the loudspeaker. The 'hump' limits the natural low frequency extension of the speaker and gives a tight, powerful - but totally false - one-note thump to the bass. The effects of stored energy do not stop there however: mid-range and treble are degraded in clarity, smoothness and information. The three-dimensional stereo image suffers badly, too.

A number of man-made materials have been used to make Hi-Fi furniture (particularly platforms), most notably Aerolam (an aluminium honeycomb), and Carbon Fibre. Both are relatively low mass and stiff but suffer from unpleasant sounding resonance and very high prices!

Recently an inexpensive Ikea table has been recommended as a turntable and equipment support. This idea is not new, having first been promoted in the mid 1980's by a turntable manufacturer of all people! To say that the table is unsuitable and acoustically undesirable is rather an understatement. The only merit I can see in it is that it is much cheaper than putting your turntable on an upturned bass drum; because that is, more or less, what it sounds like! Much as I like percussion, I really don't want a 'drum' sound effect on all the music I play. On the other hand, it might be a welcome addition to the 'orchestral', 'jazz' and 'rock' tone options on some midi systems!

Back to Nature

The most obvious material to make Hi-Fi furniture out of is wood. Its old fashioned and low tech image means it is often simply ignored in this high tech age. This is a great shame because wood is, in fact, a wonderfully versatile and 'high tech' material. Different woods have quite different engineering properties and can be used in combination to give precise functionality that no other materials can match. The engineering properties of wood have been known and understood for thousands of years. The Neolithic 'ice man' found in the Alps a few years back was carrying tools and equipment made from 29 different kinds of wood, each chosen precisely for its suitability for its exact application. That really is high tech engineering! A knowledge of the properties of different woods allows the engineer to choose the precise weight, stiffness, resonant character, bending modes, etc. to suit the purpose. For the purpose of Hi-Fi furniture, wood has several characteristics of enormous benefit. There are stiff, low mass woods available; woods with low 'Q' and high internal damping; and woods with good resonant characters. Using such woods allows us to create stiff, low mass Hi-Fi furniture.

My design for the Torlyte® Hi-Fi furniture utilises all these valuable properties of wood, including a low 'Q' broadband response, in other words the wood resonates over a wide frequency band with no one dominant resonance. What looks at first sight like simply a solid slab of wood is in fact a complex 'sandwich' containing a honeycomb inspired internal structure. More fresh air than wood, its unique combination of ultra-lightness, strength and rigidity enables it to minimise degrading acoustic feedback, and thus bring substantial sound improvements. It offers clean, extended, tuneful bass; a more realistic mid-range especially on voices; less 'splash' and more information in the treble; and improvements in the three-dimensional stereo image. The front-to-back depth snaps into focus, the recording acoustic becomes much easier to discern and those subtle, mysterious sounds made by the musicians can be identified with ease.

Whichever type of equipment stand you opt for, it's clear that you need to opt for one made of wood!

The Importance of Feet

The materials and construction used to make equipment furniture is by no means the end of the story.

Each piece of equipment needs feet and those feet play an important part in the energy 'coupling' process. Fitted with the usual rubber feet (felt and cork are also used) the equipment acts as a reservoir storing the feedback energy - and stored energy is just what we don't want! We need to dump the energy into the floor.

Luckily, energy - like water running downhill - will 'flow' into a higher mass if it can. Rubber feet or 'high tech' squidgy feet prevent this happening. They are 'decouplers' that form a higher resistance to the flow energy, trapping it in the equipment and the feet themselves.

It is often claimed that these 'decoupling rubbers' turn the energy into heat. This is a fallacy. It is extremely difficult to turn energy from one form into another and the process is very inefficient. Just how difficult it is and how little of the input energy is turned into heat is quite surprising. A few years ago I asked one of the biggest manufacturers of 'decoupling' rubbers if they could tell me exactly how much energy was turned into heat in their products. After a short silence they admitted that it was so low that they hadn't been able to measure it! That is just what I expected to hear.

Consider this for a moment: the kind of rubber we are talking about is used in squash balls. These balls are designed to bounce very little (and there are different grades or bounce rates) and to be hit very hard into a hard surface. If they convert energy into heat at any appreciable efficiency at all the heat generated by hitting the court wall would cause them to melt and run down rather than bounce off!

So what does happen to all the energy you put into the ball with the racket? In simple terms the energy has just changed down in frequency to the 'resonant frequency' of the rubber. Every material has a 'resonant frequency' the level of which is determined by its mass. Every object or substance converts energy into its resonant frequency and so stores vast amounts of energy. The low bounce rate of the squash ball is an expression of the low resonant frequency of the rubber- you can hear the 'thud'. A tennis ball, on the other hand, bounces better and you can hear the higher frequency in the 'dong' ring sound.

If you use this sort of material as equipment feet (and some people use squash balls cut in half) the feedback energy is converted down to the resonant frequency of the rubber and it just sits there and 'wobbles'. I have to tell you that I think this is a very bad idea! The destructive effect on sound quality is enormous. It increases the 'time smear' and changes the balance of the sound to a slow, boomy bass heaviness.

The Solution for Equipment

A hard material with a highish resonant frequency will perform better at 'coupling' equipment, so that energy flows quickly and easily from one thing into another and finds its way to the floor (the highest mass around) as fast as possible. The floor, of course, 'couples' the energy directly to the speaker - see Diagram 1.1 'The Hi-Fi Chain' on page 15.

A number of obvious hard materials could be (and are) used for equipment feet: steel, aluminium, titanium, stainless steel, brass, ceramics, carbon fibre, even wood. All of these materials sound different because the sonic signature of the material has changed the feedback frequency response. I have tested all of these materials and found that wood (hardwood) distorted the sound the least and its sonic character was the most 'natural'.

From my research, I have found that the cone shape is the most effective for dumping energy out of the equipment, and that the larger the diameter of the cone the better it works.

The Solution for Racks and Cabinets

If wooden cones are the best feet for 'coupling' equipment to shelves and coupling shelves to racks or cabinets, what do you use between the cabinet or rack and the floor? The answer depends on what kind of floor you have and whether or not you have a carpet.

For bare wooden, tiled or concrete floors then wooden cones are ideal. The bigger the better. If you have fitted carpet with underlay then you need feet that will penetrate the carpet to stand solidly on the concrete or wood. Machined steel spikes are the best solution, despite the fact that steel is not an ideal material in sonic terms.

If the floor under the carpet is concrete, you need nothing else, but if you have a wooden floor you need some way of stopping the spikes from sinking into the wood. Crosshead screws fixed into the floor are an excellent solution - the spikes can then be located into the screw heads (see instructions below).

Fitting spikes to a wooden floor covered with carpet.

1. Place the rack, cabinet or speaker stand exactly where you want it and mark each spike hole in the carpet. (This can be done by placing Sellotape under the spikes and pressing down so that the spikes puncture the tape.)

2. Insert a screw (use 1" (25mm) No. 8 crosshead Philips or Posidrive countersunk screws) into each hole and screw it down. The screws will disappear into the pile of the carpet and become invisible (I suggest doing this while your partner is out!). Each spike will now sit exactly on a screwhead and the result will be a great improvement in bass definition and clarity.

Why I Recommend Three Feet

Stability is a very important aspect of coupling and one that, through misunderstanding, provokes argument. One of the biggest problems with most loudspeakers is the way in which they wobble about on their floor stands or wheels. As the speaker cone moves it rocks the cabinet resulting in a loss of information and energy.

I am often asked why I recommend using three feet. The answer is simple: stability is vital and whilst four feet may look more solid they allow micro-rocking. Even if the rocking action is very small, in this situation (dumping energy) it is very significant. By using three feet (two placed at the front, one at the back) you create a stable triangle and prevent information loss. Unconvinced? The improvement brought by using three points rather than four can easily be shown by doing a simple listening test.

The following diagrams help demonstrate this point:

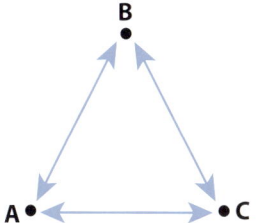

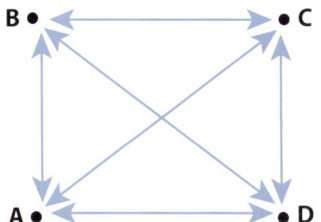

Diagram 4.7: Three points of contact, A, B and C, give stable balanced, triangulation of vector forces. No rocking will occur, even if one point is higher or lower than the others. That's why camera tripods have three legs: stability is as crucial for taking a photo as it is for reproducing music!

Diagram 4.8: Four points create four triangles of force: ABC, ABD, ACD, BCD. The object can, and will, rock from one to another. Even if this rocking is very small, the impact on the reproduced sound is large.

Of course, three points of contact on a square or rectangular object mean that you have two unsupported corners that can be pressed down to tip it over. The feedback energy isn't able to press down on one corner only so that isn't a problem; it's just a problem if you lean on an unsupported corner! That is a different kind of stability and one not necessary in this situation. For aesthetics and practical use I recommend putting a cone under each front corner and one in the middle at the back.

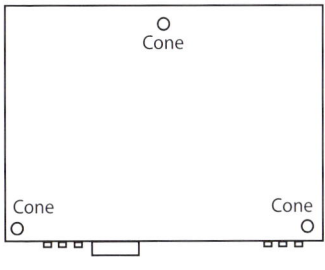

Front of Equipment

Diagram 4.9: Placement of three feet under equipment (two at front corners, one centre back).

The Floor - The Final Foundation

The floor of the listening room is an integral part of the acoustic feedback loop and as such must be carefully examined. As many of you will have discovered, floors can have a major degrading effect on the sound of your Hi-Fi System. The most common problem is the bouncy wooden floor, practical solutions to which I've addressed in the next chapter .

If you are unable to implement the solutions to bouncy wooden floors, you could consider wall mounting your system on shelves - but only if you have a solid wall! This will also be the solution if you are unable to cure your floor problems using the methods suggested there. If you live in a rented flat or house, sorry but there's very little you can do, except buy a house or headphones - whichever you can afford!

Chapter Five
Shake, Rattle & Roll

STEP 5

Upgrade Path Step 5: Room Acoustics

Let me make one thing quite clear from the outset; the advice in this chapter is for domestic use not professional studios. That is why I have not given it the title 'Room Acoustics' though that is, in its sense of acoustics meaning 'sounds', what it is all about. This chapter exists because the sound of the room (or the sounds in the room) has such a profound effect on your Hi-Fi system. It also, of course, has a marked effect on your appreciation of the performance of your upgrades.

The factors that have an influence on the sound and your enjoyment of your system include the position and placement of your system and speakers; the size and construction of your room; the way the room and furnishings absorb the sound 'echo' within it and the willingness of your room to 'sing-along' with your system. The emphasis here is to show you what you can do to improve the sound of the room by looking at these factors, finding out where the problems are and then fixing them. You don't have to do any of the structural work yourself, use professionals whenever appropriate.

As I have explained in detail in Chapter One, a Hi-Fi system is not a simple chain with an input (the source) at one end and an output (the loudspeakers) at the other. It is a complex interactive loop with many points of feedback where it interacts with the room, its supports and itself.

Most advice on room acoustics - whether it is a magazine article or textbook - includes maths for calculating room eigentones, reverberation times, absorption coefficients, etc. This sort of technical detail, interesting though it is to the scientifically inquisitive, is not of any practical value in a book like this. From time to time there are also several 'consultants' advertising in Hi-Fi magazines offering and providing 'cures' with special 'room tuning' devices. No amount of this kind of analysis can tell you about the benefits of Blu-Tack™ or identify problems such as four wobbly points of contact, loose driver screws or poor speaker placement. No amount of frequency-tuned absorbers can possibly solve these problems.

The room has a Sound?

You may have noticed that your system sounds different in different places in the same room and also in different rooms. The effect of the room 'sound' on the sound of a Hi-Fi system is enormous and, in Hi-Fi circles, badly misunderstood. In audio scientific circles it has been researched and measured but not explained. In 1992, four Canadian scientists presented a paper to the 93rd Convention of the Audio Engineering Society, of which I am a member, on 'The Effect of Loudspeaker Placement on Listening Preference Ratings'. An extract states:

"Room placement of a loudspeaker measurably affects listeners' judgement of its quality, and some changes in positioning are more easily perceived than the substitution of a different loudspeaker. Speaker placement similarly modifies quality judgement of the programme material".

No explanation of these effects was offered.

I believe that the explanation is quite obvious. As discussed in Chapter One, the room and the loudspeaker (plus the rest of the Hi-Fi system) are interactive. Each has profound effects on the other and to understand what is happening in one direction, you have to know what is happening in the other. It is important to appreciate that not only does the loudspeaker feed a signal into the room, but that the room feeds that signal plus the room sound back into the loudspeaker (see 'The Hi-Fi Chain' on page 15). The loudspeaker acts as a very efficient microphone at the same time as it is working as a loudspeaker!

The problem is rather like that of designing a good car suspension. The suspension exists to even out the road surface imperfections and to allow safe cornering, etc., but its performance is complicated by the car body sitting on top of it. The road is moving one end of the spring and damper, whilst the body is moving the other end - each in response to what the other is doing. Automotive engineers understand this interactive relationship but, bizarrely, most audio engineers do not understand the relationship between a loudspeaker and the room. It is clear that they don't consider the relationship because they continue to design loudspeakers to give an even, flat frequency response (the usual design criterion) under what is called 'free field' conditions. 'Free field' means away from any reflective surface, i.e. up a 20 foot pole in a field!

Since this is a very inconvenient method - rain, snow, wind noise, etc. - the Anechoic chamber was invented as an indoor substitute. Nearly every Hi-Fi loudspeaker is designed and tested this way.

It's as if the car was designed and road tested only on a perfectly smooth, flat, straight road with no hard acceleration, braking or turning involved. You can imagine what would happen to such cars when driven on real roads under real driving conditions! The car would be very hard to control, bouncing all over the place, lurching into and out of corners and dangerously unstable.

Yet this is exactly what we are expected to put up with in a Hi-Fi system when loudspeakers are designed with loose, floppy drive units and complex crossover networks. Tragically it is the more expensive, larger 'esoteric' loudspeakers that are the worst in this respect. They are designed to give fuller, deeper bass - right in the frequency area where the room has most effect on them.

This isn't meant to be a diatribe against Hi-Fi loudspeakers designers or manufacturers, but an explanation of what is going on when you play a Hi-Fi system in a room. The room/loudspeaker interaction is only one of the problems as you will discover when you read on. My purpose is to help you to recognise the problems and solve them in an inexpensive and effective way.

What Influences the Sound of a Room?

We don't need special test equipment to tell us how good or bad a room is. We have ears and we can tell a lot about a room simply by talking in it.

A trick like snapping the fingers or clapping can be useful for revealing the high frequency echo qualities, but speech is all we need to judge the sound qualities of a room. Those of us able to play or listen to a musical instrument being played in a room will have noticed how few rooms are 'bad' to listen to live instruments in compared with how many are 'bad' to listen to Hi-Fi systems in. That is because an instrument is a true 'source' and interacts very little with the room. You can still hear the room's character, however, on speech or with a live instrument.

That 'character' may be improved to the benefit of both live and reproduced sounds in ways that are inexpensive, effective and aesthetically acceptable.

Room Proportion and Low Frequency Modes

All the books and articles on room acoustics that I have read, whilst being (for the most part) technically correct, make an important but unstated assumption. They talk about room dimensions and the associated room modes, eigentones (room standing wave frequencies defined by the distance between wall, ceiling and floor) and low frequency responses, assuming that all the walls, the floor and the ceiling are totally reflective at all frequencies.

They also assume a totally closed box! Real rooms aren't like this and differ from the theoretical model in a number of very important ways. These differences seriously complicate and confuse the simple model the authors propose. For example: a simple aperture like a doorway or window in a room creates a Helmholtz Resonator (an acoustical resonator which is sharply resonant at a single frquency); a large picture window, French window or patio door turns the room 'box' model into an open pipe model (basically a 'tube' where the air in the pipe resonates at a frequency equal to the velocity of sound divided by twice the length of the pipe. All the harmonics of that frequency are also present). Each of these have distinct and different acoustic properties.

Real domestic rooms are rarely constructed with all surfaces reflective at low frequencies. Solid brick, stone or concrete block outside walls are reflective to low frequencies because their mass is high enough, but light stud or partition walls are not. Light partition walls are absorbers of low frequencies because they flex. Low frequencies travel right through them to be reflected off the nearest solid outside wall.

The low frequencies generated by the loudspeakers in your listening room may, therefore, 'see' only the solid outer walls of your house. These are the dimensions that determine the low frequency eigentones or modes, not the walls you see in the room!

Calculations for the height are similarly misleading because the 'floor' will not be the wooden suspended floor but the concrete raft beneath. The ceiling will also be of too light a construction to form a low frequency reflective surface. Low frequencies may be reflected from the roof rather than the apparent ceiling. The room you see is not necessarily the room that your loudspeakers 'see'. The lowest frequencies in the music may only be heard properly outside the room.

My own room illustrates this point. It is on the top floor of a converted stone barn. I have very large loudspeakers with a 'healthy' very low frequency output and I can hear all the deep bass within the room. The problem is that I can't hear the lowest frequencies at the sound level they should be to balance with the rest because the modes fall outside the actual listening room.

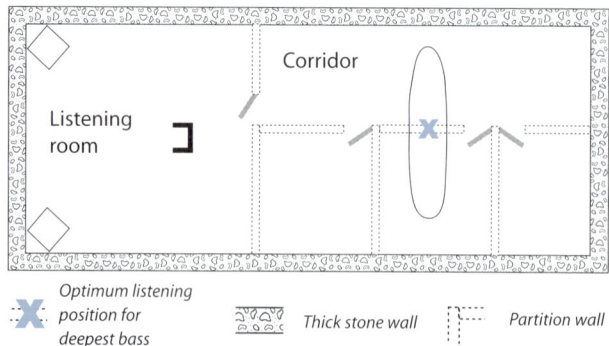

Diagram 5.1 My room layout

For those frequencies, the gable ends define the room length, rather than the partition wall that divides off my listening room. Another way of putting it is that for those low frequencies I'm sitting much too close to the speakers although in the right position for all the other frequencies. It should be realised that this isn't necessarily a bad thing to happen if your speakers have loose, sloppy uncontrolled bass.

On the other hand, I experienced the downside of this in the early 1970s when I installed a pair of B&W DM70 loudspeakers for a customer in a block of new flats. We were standing scratching our heads about the lack of deep bass when there was a knock at the door. It was a flat owner from down the corridor. He just said 'Whatever you are doing, please stop it because it is making me feel sick!' We trooped down the corridor to his flat and found my customer's bass - and boy was it nauseating!

It is obvious, therefore, that only actual measurements can be relied upon and calculations from textbook formulae can be misleading.

System Placement

To avoid repetition, I will assume that you have read the previous chapter and are familiar with the acoustic feedback and time delay problems and my solutions. (If you need to refresh your memory, turn to page 75.)

Here, I will deal in detail with the room and what you can do to minimise the effects of feedback and get the best sound of which your system and room are capable. System and loudspeaker placement is critical but a great deal of misinformation is given by 'experts' and Hi-Fi magazines. The two most common misconceptions are to keep your speakers well away from walls and to put them close together to avoid the so called 'hole in the middle'.

Myth 1: Keep your Speakers away from Walls

The exact position for a loudspeaker in your room will depend on the design and size of the loudspeaker and on the shape of your room. Some designs need to be very close to the wall, others need to be up to three feet from the back wall.

Large loudspeakers with big floppy bass units may react badly to being placed in corners and large Quad electrostatics will not give their best sound if used close to a back wall or in a small room. It is up to you to experiment within the guidelines I have indicated in **Perfect Speaker Positioning** on page 93. You cannot rely on the advice given by the loudspeaker manufacturer without testing it for yourself.

Myth 2: Put your Speakers Together to avoid a 'Hole in the Middle'

Most speaker manufacturers and magazines say that it is best to form an equilateral triangle between the speakers and your seating position.

They show a diagram like the one below that suggests that the speakers should be quite close together and you close to them. The result of this arrangement is a very small, intense and confused soundstage. This is as unsatisfactory and unconvincing as running a fancy Home Cinema system using a 14" TV set as the picture monitor!

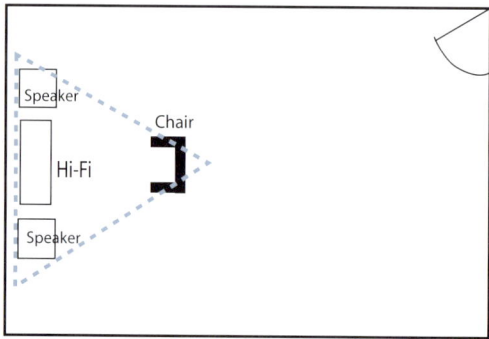

Diagram 5.2: The 'ideal' set up, according to most speaker manufacturers and Hi-Fi magazines.

The advice in a popular Hi-Fi magazine that, "*If the distance from you to each speaker is less than the distance between the speakers, the sonic 'picture' created will have a hole in the middle*", is complete nonsense! It is absolutely impossible for a hole to appear in the middle. Think about it; if this were the case, headphones wouldn't work. What you are trying to achieve is the correct angular and depth perspectives of the sound sources.

Site your speakers as I suggest and it will open out the soundstage to the size of the back wall. It is much more realistic, the separation between instruments and performers is greater and the 3-dimensionality is very much deeper.

You will see from the illustration of my preferred speaker layout that I have placed the equipment on a rack or cabinet between the loudspeakers. There are a number of advantages to this arrangement: it keeps the acoustic feedback path short between the speakers and the equipment; it allows for one multisocket outlet for the Hi-Fi mains ring circuit; and it means that if you have floor bounce problems you can restrict the treatment and therefore the cost and inconvenience.

Perfect Speaker Positioning

1. Place the speakers as wide apart as they can go but no less than about 2" (50mm) from the side walls and close to the back wall. In a rectangular room, choose one of the short walls with no doors or obstructions to prevent you placing your equipment in the centre or running the speaker cables. Choose a solid wall rather than a light partition wall or one with a French window, a bay window or a door. Not only is keeping the equipment rack in the centre tidy and convenient, it keeps speaker cable lengths shorter and of equal length. Good loudspeaker cable is expensive, so for a given budget the shorter the cable the better it can be. But don't compromise on speaker separation – it's worth more than better cable.

2. Move your speakers forward a little at a time to find the right balance between bass tightness, depth and midrange colouration.

3. Angle the speakers towards the listening position (toe-in) to increase the focus and solidity of the image. It gives you a greater ratio of direct to reflected sound.

4. Place your listening seat in the preferred listening area and move it forward or back to find the soundstage 'sweet spot' where the image comes into focus giving believable positioning and separation. Live recordings are usually better for this purpose than studio recordings.

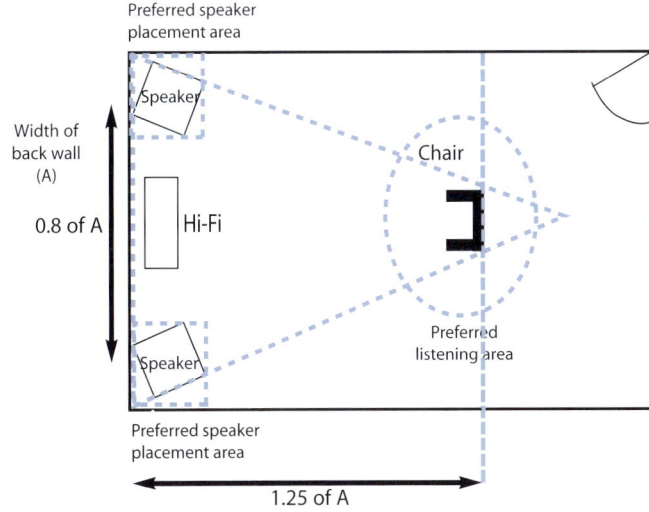

Diagram 5.3: My preferred set-up

Room Dimensions

Every room has several low frequency resonances caused by its dimensions. There are three dominant ones defined by the height, length and width of the room and each has 2nd, 3rd, 4th, etc. harmonics. You can't stop them occurring but you can choose to use a room where they don't coincide and become a serious impediment to musical enjoyment.

If you are thinking about building a music room then, of course, you can choose to build it to the Golden Ratio. This ratio is different depending on whether it is a small room or a large room.

If I take the example of a small room, if the height is the old standard British value of 8ft (2.44m), then the width should be 10ft (3.05m) and the length 12'9" (3.89m). Fortunately, this is a very common room size so most people have little to worry about.

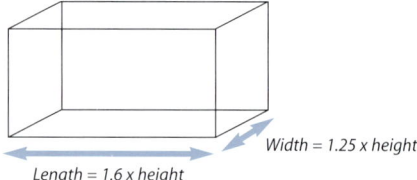

Width = 1.25 x height

Length = 1.6 x height

Diagram 5.4:
Golden Ratio for a Small room

For a large room with an 8ft (2.44m) ceiling, the width should be around 12'9" (3.89m) and the length 20' (6.1m). Again, this is a very common size.

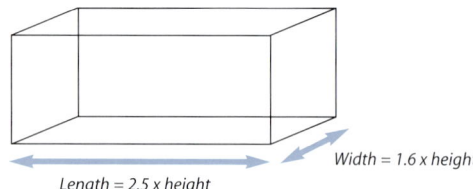

Diagram 5.5:
Golden Ratio for a Large room

Width = 1.6 x height

Length = 2.5 x height

You are only likely to run into problems if two (or even three) of your room dimensions or multiples of them coincide and so the resonances add up. It is possible to design resonant absorbers to control specific problem frequencies but my experience of these used in domestic environments is that they tend to cause more audible problems than they solve. I believe that they are sensible engineering solutions for use only in carefully designed and controlled professional environments like recording studios. Generally they also have the domestic disadvantages of being both ugly and expensive.

Russ's Tips
Make sure your speaker drive units are secure

Before you start repositioning your speakers, take the front grill off, get screwdrivers or spanners of the correct size and tighten up all the screws or bolts holding the drive units into the baffle. Don't overtighten - just 'nip-up'. Do the same for any screws that secure the front or back baffle board. Whether your speakers are brand new or old friends, I guarantee they'll need tightening and that you will be amazed by the improvements. The bass will be much tighter and cleaner, there will be more information in the midrange and the treble will be sweeter.

Fit the correct feet to your speakers

Don't waste your time fine-tuning the precisely perfect spots for your speakers unless you have followed the cone/spike advice given in the previous chapter (page 84) on feet. That alone will have such a large effect on the sound that you should get it sorted before putting too much work into finding the best speaker position. Again it's important to take the right steps in the right order to avoid confusion and wasted effort.

Recording studios are hopeless as domestic music rooms and vice versa. It is a mistake to use the solutions of one to solve the problems of the other. I will be recommending room tuning measures later on in this chapter, but they are specifically designed for domestic use and are things you can make yourself. If, however, after putting into practice all of my advice here and previously, you find that you have a severe room resonance problem that needs a more drastic cure, then by all means try out the resonant absorber solution. But first identify exactly what the cause is. It is a solution to be tried last not first!

Remember, it is the interaction between your system (especially your loudspeakers) and the room that is the problem rather than just the room itself. People often notice that a given system sounds different in different rooms and then jump to the conclusion that the only variable (and therefore the culprit) is the room. The truth is that some rooms will be kind to the system by not showing up the problems and others will be cruel by highlighting them. If the reality is interaction between system and room, then the only rational, productive route to take is to work on them both.

The Room Structure: Traditional Room Construction

The typical British house built during the 20th Century is a solid, double brick outer wall construction with brick internal walls. Typically, it is two storey, with wooden floors supported on joists. Roofs are pitched and hung with slate. Ground floor internal walls are usually solid and plastered, though some first floor walls may be light partition or studding. In some areas (South and West England) single storey bungalows predominate and in others (towns and cities) multistorey flats, apartments and tenements are common, built in brick, stone or concrete.

These construction methods, though producing different sound characters, present few problems and generally sound acceptable. The exception is the 'concrete box' construction of modern flats or apartments, where 'flutter' echoes (page 105) can be quite troublesome (see **Modern Room Construction**, page 99).

Floor Problems

The main structural problem encountered by the Hi-Fi enthusiast in the traditional house is bouncy wooden floors. Though originally probably very solid, the wooden wedges and packing pieces will have deteriorated with age and fungal or insect attack. The floor 'boom' and colouration feeds back into the loudspeaker, back to the amplifier where it is further amplified in a classic feedback loop. The solution to the problem is simpler and easier than you might expect, but will involve lifting floorboards and crawling about in the space under the floor. Better to get a competent builder or joiner to do this simple job for you. There follows details about what you need to do depending on the construction of your floor in the blue information boxes on pages 97 & 98.

Walls

Plastered walls give few problems, though there are a couple of exceptions: loose areas of plaster, blocked up windows or doors, and flutter echoes.

Loose areas of plaster held in place by wallpaper, will rattle in sympathy with low frequencies and produce some very strange sounds! They must be cut out and re-plastered. The new plaster must be carefully chosen to match the strength and hardness of the original or the new areas will sound quite different from the original. The temptation is to fill the repair with one layer of all purpose hard plaster, but don't let your plasterer do this. Make sure he fills the repairs in layers just like the original: a 'sand and cement' or 'browning' layer, and a harder 'skimming' layer.

Floor boards or flooring chipboard (Weyroc) nailed to joists

If yours is a ground floor room, the usual cause of bounciness is that the joists have become unsupported at their centres - and sometimes at the ends.

It is usually necessary to remove a few floorboards (or small areas of flooring chipboard) to allow access to the underfloor space. Replace the wedges supporting the centre of the joist and/or pack out the ends of the joists if they are 'floating'. It is usually only a couple of hours work for a competent builder. When the joists are solid again, screw down the floorboards/chipboard in the area where your Hi-Fi system is situated. This will eliminate board 'rattle'. You may need to do this elsewhere if other boards are no longer tight to the joists.

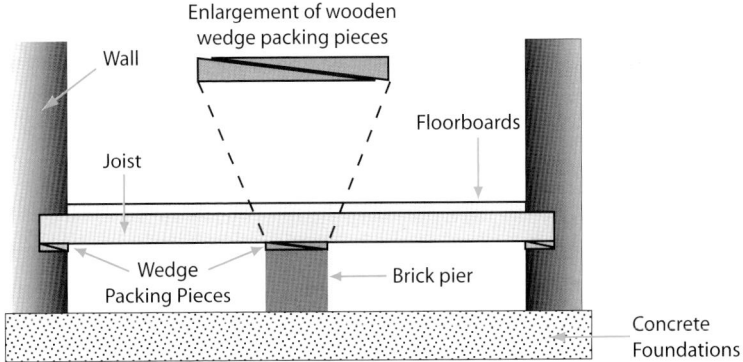

Diagram 5.6: Ground floor room with wooden floor

If you have a ground floor with a cellar or basement beneath it, you can support the rafters with wooden pillars. Use at least 3"x 4" (76mm x 102mm)section timber. Cut the pillar short and use the overlapping wedge technique to take out the joist bounce without putting it into tension.

Suspended Floors

A common problem area, suspended floors can be stiffened by fitting 'Herring-bone Strutting' between each joist. The strutting is short lengths of 3" x 1" or 3" x 2" (76mm x 25mm or 76mm x 51mm) nailed diagonally between and to the joists spaced about 3ft (1m) apart.

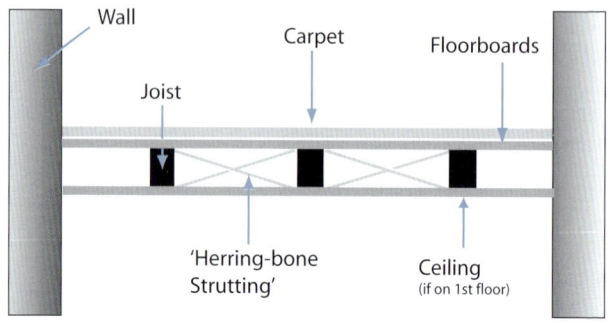

Diagram 5.7: Suspended floor with 'Herringbone Strutting'

'Floating' floor on concrete

This type of floor can cause problems (i) if the battens supporting the chipboard floor are not secure to the concrete, or (ii) if the chipboard is not well nailed to the battens.

Ideally, the battens should be nailed down with masonry nails or screwed down to the concrete. The chipboard floor panels should, in turn, be screwed to the battens.

If your floor is a real problem, you might consider lifting the boards and refixing the battens. You could also put down extra battens - say at 18" (450mm) spacings. To save time and disruption you could just do this at the system end of the room.

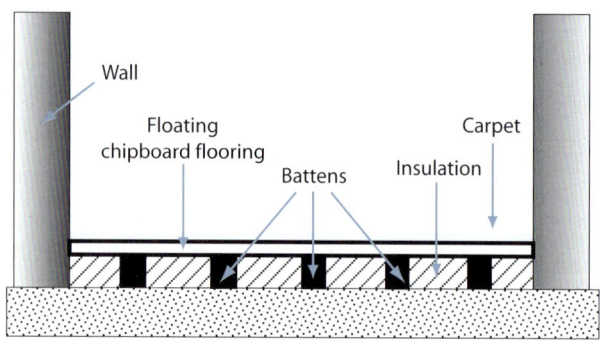

Diagram 5.8: Ground floor room with floating floor

Builders will normally use concrete blocks and a simple thin, hard, modern plaster finish when blocking up window and door openings in old brick or stone walls. The results can sound disastrous, but you can solve the problem using wall hangings and/or insulation board over the offending area.

I was once called to the home of a customer who complained that his Hi-Fi system had suddenly become bright and very 'hard' sounding. His house was solidly built of dressed stone with thick, soft old plaster on the walls.

The 'sound' of the room was excellent in most places but I identified (by simply clapping my hands) that an area of wall right beside his listening seat sounded quite different from the rest. There was a nasty 'flutter' echo from it and general hardness. When I said "There's something strange about this area of wall - have you bricked up a window?" he thought I had some kind of special powers to see through wallpaper! He had indeed just had a window filled in and the room redecorated!

He was understandably reluctant to redo the work using Thermalite block and 'soft' plaster but I persuaded him to hang a decorative wall hanging backed by a sheet of insulation board over the ex-window area and hide the problem. The 'cure' worked I'm pleased to say! For more information on these type of echoes caused by modern plaster, see the section on 'Flutter Echoes', page 108.

The Room Structure: Modern Room Construction

In response to rising standards of thermal insulation in the Building Regulations, and the demand for faster, cheaper construction methods, a wide variety of techniques are now in use. The relatively modern double brick cavity wall led to brick/concrete block, concrete block/concrete block and concrete block/Thermalite block constructions that have now been replaced by single wall construction using new techniques and materials. Wood frame construction is becoming more common with either timber or brick cladding. Wood frame houses are very common in other countries where, perhaps, low cost and speed of construction have been valued higher than their 'traditional' housebuilding methods of bricklaying and plastering.

Each building method has its own acoustic signature that you must accept because you can't change it. Of course, the sensible approach is to put the needs of Hi-Fi reproduction at the top of your priority list when looking for a new house! Audition each sitting room and assess it for practicality in terms of door and window placement in relation to system placement and listening position.

Concrete Block Cavity Walls, and Concrete Outer/Thermalite Inner Leaf Cavity Walls

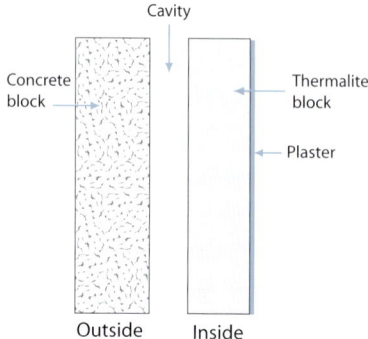

Diagram 5.9: Cross section of a concrete outer/thermalite inner leaf cavity wall.

These are merely modern versions of the traditional construction. The difference is in the acoustic character of the materials, concrete sounding much 'harder' with a more obvious nasty 'zing' to it than brick or Thermalite.

A 'concrete box' type room is very unkind to Hi-Fi systems and the more powerful they are and the bigger the speakers, the worse they sound! The secret is to use small speakers of modest power but with a very smooth (almost dull) response. The less energy you put into the room the better the sound. This principle is extremely valuable at Hi-Fi shows held in modern 'concrete box' hotels. Experienced 'old hand' exhibitors use small speakers and produce much better sounds than the enthusiastic 'young tigers' who try and impress with big speakers and mega-watt amplification.

Dry-lined Walls

Often outside walls are 'dry-lined' with a wooden frame nailed to the inner block with special plaster boards nailed to the frame or more recently the plasterboards are just glued directly on the Thermalite blocks. The gaps between the boards and the nail holes are then 'repaired' with plaster and simply sanded flat. The surface is then ready for painting or wallpapering. This construction is quicker, easier and cheaper for the builder because it reduces construction time, but the acoustic results can be very variable.

The variability is caused by the relative complexity of the structure. The plasterboard on the framing acts as a 'panel resonant absorber' that absorbs low frequencies, and the gaps behind the plaster board act as 'cavity resonators' (and to really confuse the situation they rattle if not very firmly fixed). The effect of the '

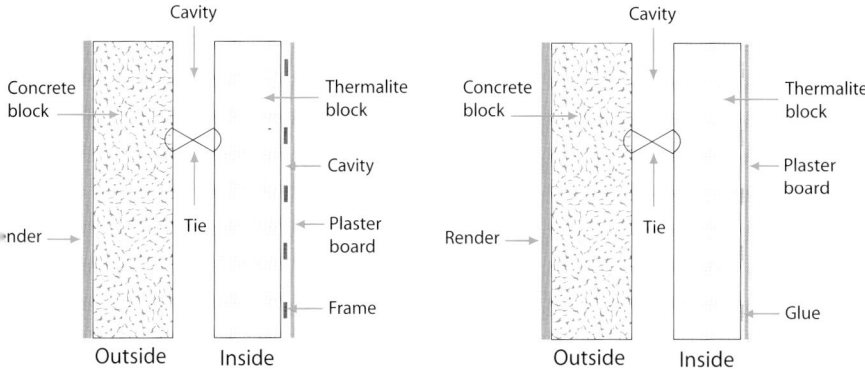

Diagram 5.10:
Cross section of a dry lined wall using frame.

Diagram 5.11:
Cross section of a dry lined wall with
plasterboard glued directly to blocks.

panel absorber' may or may not be a problem depending on the loudspeaker used. The loss of bass may be too much for small speakers, but it may be quite effective to control the bass boom of some large speakers. This is a speaker and room interaction problem that tends to dictate your choice of loudspeaker!

The problem can be reduced considerably, however, with the injection of structural foam into the cavities between the plasterboard and the Thermalite block. See instructions for filling the wall cavity with foam on page 105. You may wish to wait, though, until you need to redecorate before tackling this upgrade, as you certainly will need to afterwards! (I advise caution here, because some of current 'dry lining' techniques and materials could be compromised by this method. There may, for example, be a vapour barrier or insulation that should not be penetrated, so seek the advice of an architect or good builder first. If you are buying a new house under construction, a few quid to the foreman or joiner to make sure the framing is fixed more solidly would pay dividends!)

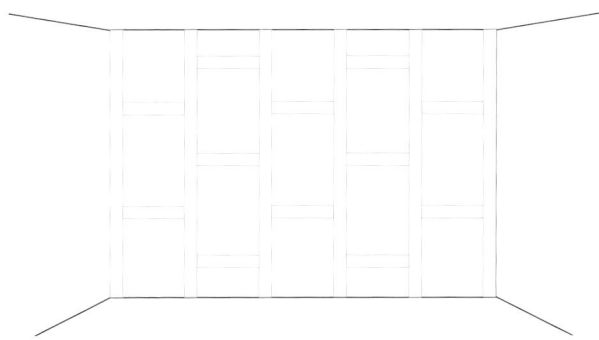

Diagram 5.12: The frame and cavities of a dry lined wall

Instructions for filling the wall cavity with foam

You will need to buy yourself a joist detector and an electrical wiring detector. These are usually combined in one unit. You will also need a supply of large cans of structural expanding polyurethane foam filler and sealer (obtainable from DIY stores or builders merchants).

Using the 'joist' detector function, map out on the wall with a pencil the positions of the uprights and horizontals behind the plaster board. Then do the same for the electrical wiring using the wiring function. Before drilling any holes, switch off the circuit breakers to any wiring found (check again with the detectors).

Your wall 'map' will show you where all the cavities are behind the plasterboard. Now drill a hole big enough to insert the foam applicator tube (approximately 9mm diameter) in the centre of each cavity. You'll need a battery powered drill - remember you've turned off the power! Have some masking tape handy to seal each hole after using the foam, as it expands 2-3 times its volume and will ooze out of the hole again like an alien monster unless you seal it.

Squirt foam into each of the holes, putting in enough to glue the panel to the block wall behind. Use your own judgement on this - too much won't matter (except for the cost) but too little is a waste of time and effort. Remember that it expands 2-3 times its volume a few seconds after coming out of the nozzle, so I suggest that you test the expansion dynamics first on a sheet of old newspaper. Read the instructions on the can! When the foam has cured, test the wall for rattles by banging it with your fist (like you do in the Police Station or pub to get some service!).

If the wall still has rattles where the uprights are insecurely fixed to the blocks, drill further holes next to the upright where it is rattling and inject some more foam. After the foam has cured completely, you can remove the masking tape and sand the foam flat to the wall. Redecorating may now begin!

IMPORTANT NOTE: Check, of course, that there is no fibreglass damping material already filling the cavities. Before launching into any of this foam injection, do consider the effect of the dry lined wall or partition wall as a 'resonant bass absorber'. Some of that bass absorption may be beneficial although usually it is the cause of lack of bass in rooms.

Partition (or Stud) walls

They are simply built of studding (typically 2"x4" (51mm x 102mm) timber) clad in plasterboard so the same expanding foam technique as for dry-lined walls can be used to stop vibration rattles and limit low frequency absorption. Sound transmission through the wall will be reduced considerably if you fill each cavity rather than use just enough to damp the centres.

Sound Absorption

This is the amount of sound 'echo' damping within a room - not to be confused with 'sound proofing' which is a technique for stopping sound getting out of a room. Soundproofing is a big technical subject in its own right and outside the scope of this book.

In order to achieve an acceptable reverberation time (i.e. good sounding), a room needs to contain a certain amount of absorptive materials. Fortunately, a normal sitting room with a carpet, curtains, upholstered suite and other furnishings usually achieves an acceptable reverberation time. Modern minimalist rooms with lots of glass are a real problem that I am still working on.

Obviously, the enthusiast can fine-tune his room's baseline acoustics by adjusting the details of those basic furnishing items. The human body is sound absorbent too (there's a joke in there somewhere!) so don't forget that the listener's presence is affecting the room acoustics. I have been in a number of Hi-Fi listening rooms that were too live in the mid range until four or five people were present. For a solo listener these rooms needed more 'mass' absorption.

The extra absorption can be achieved in a number of 'invisible' ways that can broaden the absorption band into the high frequency area to great advantage. My aim is to use materials and techniques that achieve smooth broad band absorption rather than narrow band 'problem' curing solutions.

Our ears are very sensitive to the characteristics of rooms so that we feel uncomfortable in rooms with large variations (anomalies) in the frequency response and reverberation times.

I have found, over the years, that manmade materials tend to have narrow-band effects but natural materials have broadband responses. I therefore try to avoid using foam, rubber or plastic materials as dampers, but these days soft furnishings are only stuffed with foam. Old chairs or settees with horse hair and kapok, and cushions stuffed with feathers sound much better in a listening room. Leather upholstery sounds better than vinyl, and I suspect that leather also masks much of the negative effect of the foam filling. Layers of Dacron over the foam help too.

Narrow band effects and acoustic anomalies are caused by several related factors. The narrowness of the band itself is clearly audible and is therefore noticeable in the overall sound. The real problem, however, is in the phase anomalies created at the start and end of the narrow band. They are associated with any rapid rise or fall in a response curve and the ear is very sensitive to them and so notices the anomaly.

Carpet and carpet underlay

Carpets can have a profound effect on the sound of a room. A fully fitted carpet covers a large proportion of the surface area of a room and so exerts a marked effect.

It is possible to overdo the damping effect with a wall to wall carpet, though rarely, but bear it in mind. In general terms a thick pile carpet (especially of high wool content) is much better sounding as it absorbs over a broader band than a thin manmade Nylon fibre one. Obviously, the carpet to avoid is the cheap nylon rubber backed type - it sounds terrible!

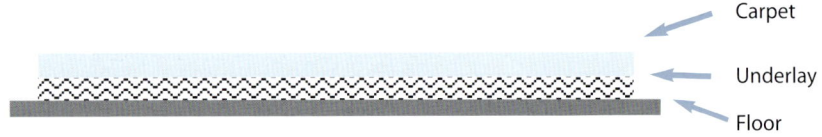

Carpet

Underlay

Floor

Diagram 5.13: Felt underlay is better than rubber

Carpet underlays are clearly audible in their effects too, so avoid the ubiquitous rubber underlays that shops now expect you to want, and choose good old-fashioned underfelt… and buy the best quality. Felt underlay improves the low frequency absorption of the carpet. It feels better to walk on than the rubber too.

You can vary or tune the effect of carpeting by using rugs on top of a fitted carpet, especially in the area between the listener and the speakers. A large good quality rug, such as Indian or Chinese carpet placed in the centre of a room on an existing carpet can be extremely successful.

Alternatively it could be laid on stained and polished floorboards, parquet flooring or with a thin carpet surround. (When surrounded by polished wood I would recommend two layers of best quality underfelt.) The best sounding rooms I have had have used this 'layered' technique.

Curtains and Windows

Curtains are very useful indeed in controlling delay time and absorbing mid and high frequencies. They are more effective if they go right down to the floor and are lined.

A thick material like velvet or velour is most effective, and you can make them more effective still if you allow extra widths of material so that the folds are deep. The deeper the folds, the more mid range is absorbed. It is important to always line curtains. If a thin curtain material is preferred, an interlining of Dacron could be used.

Velvet curtain

Interlining

Lining

Diagram 5.14: Thick curtains will absorb mid and high frequencies

Windows absorb quite large amounts of bass because both the glass itself and the flexing of the glass absorb bass. When you draw a curtain across a window you create a complicated kind of diaphragm absorber using the window recess as a 'cavity absorber' and the window glass as a 'resonant absorber'. It's no wonder that the room sound can change dramatically when you draw the curtains!

Reflections and Flutter Echoes

Having dealt with the most influential area of furniture, carpets and curtains, we can turn our attention to the fine-tuning details. The walls and ceiling of a room are large reflective surfaces and the strength and frequency band of reflection greatly influence things like spaciousness, depth of image, image stability, width, etc.

Where sound bounces repeatedly back and forth between reflective surfaces, the result is 'flutter echos' and even more sound degradation.

Reflections

In your listening seat you hear sound from two sources. You hear the direct sound from the speakers first, followed by the reflected sound from the floor, ceiling and walls. The strength of the reflections and the time delay after the direct sound govern the 'spaciousness' of the room sound. This room sound overlays the recorded acoustic and so must be very carefully controlled. It is all too easy to overdamp a room and produce a closed-in claustrophobic sound.

You can avoid this pitfall by being very selective about the placement of absorbers and by creating a relatively 'dead' end where the speakers stand and a relatively 'live' end where you sit; though strong reflections from the walls near your seat are very distracting and uncomfortable. If the speaker end is 'live', and you sit in the 'dead' end, the sound is too distant and echoing. If the speaker end is 'dead' you hear a more forward direct sound with the room acoustic added - a much more natural and satisfying sound. Of course, this only applies to a two-channel system.

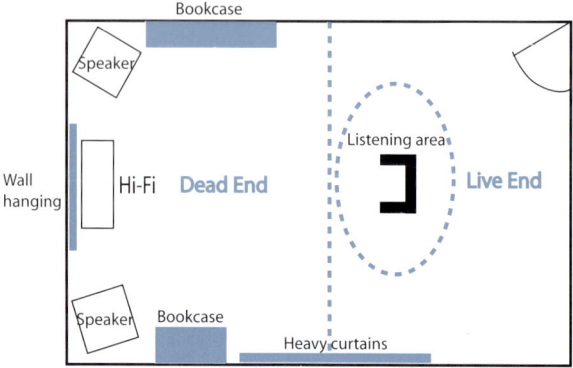

Diagram 5.15: A listening room, showing the Dead End, Live End and Absorbers

Before using any absorbent wall treatment however, it is essential that you place the loudspeakers and your listening chair in the best positions. Many reflection problems are simply eliminated this way.

'Dead End'

Conventional advice on the subject of speaker placement, *"loudspeakers should be located as far away from reflecting surfaces as practicable"*, (F. Alton Everest - The Master Handbook of Acoustics, 3rd Edition, page 341) is guaranteed to increase the problems of reflected sound (amongst others!). This technique increases the distance the sound must travel and so increases the time delay. It may, superficially, enhance spaciousness in a system where the speakers are too close together; but two wrongs don't make a right.

Everest (and others) attempt to solve this problem by placing absorbers on all reflecting wall and ceiling surfaces. This modifies the reflections, producing numerous audible anomalies and puts the 'dead' end at the wrong end of the room. In my experience it also 'overdamps' the room and tends to make it look and sound like a recording studio. I feel strongly that a domestic listening room should remain looking 'domestic' and that acoustic treatment should be modest and 'invisible' as such.

A better solution to the reflection problem is achieved by placing the speakers as wide apart as is practically possible - closer to the rear wall and corners than is 'fashionable'. This technique places the sound source close to the first reflective surface, shortening the reflection path to the point where it is indistinguishable from the source itself.

It also moves the main reflective area further down the room where it is easily dealt with by the room furnishings (chairs, curtains, bookcases, record cabinets, etc.) and the absorptive treatments. 'Toed-in' to fire directly at the listening seat which should be as far back as is practically possible.

The role of room furnishings to absorb and diffuse reflections

Bookcases and record storage cabinets make excellent sound diffusors and absorbers. A diffusor breaks up reflected sound waves by being an uneven surface. An absorber breaks up reflected sounds by absorbing them in the material it is made of. A bookcase (without doors!) combines both types because the books themselves absorb sound and their arrangement and uneven size diffuses the sound. Record storage works similarly but not quite as effectively.

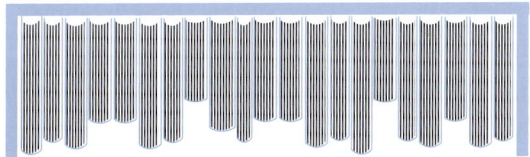

Diagram 5.16: View of a bookcase from above

Your record storage unit could, quite conveniently, be placed to the side of the listening chair and a bookcase (or extra record storage) placed on the wall behind. Unfortunately, CD storage won't be much help here, another good reason for owning a record collection! Reflections from behind are clearly as distracting as those from the side. You will find it quite easy to listen for the reflections and then deal with each in turn.

Coffee tables in front of the listening chair are often very audible and degrading. The top forms a reflective surface in just the wrong place! Better to put it to the side of your chair in the room.

Flutter Echoes

'Flutter echoes' occur between hard parallel walls where the sound repeatedly bounces back and forth. They sound very nasty indeed and must be cured. It is very easy to identify where they occur by simply clapping your hands. Walk up and down the room to locate flutter echo points and mark them for treatment.

You only need to treat one wall to cure a 'flutter' so you can choose which one to suit the decor or furnishing arrangement. Flutter echoes can occur in just one place, or can extend right down the whole length of the room.

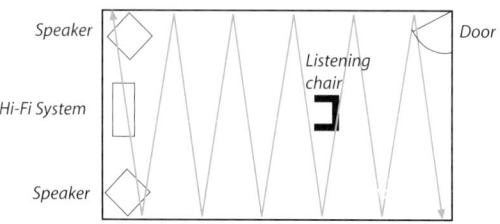

Diagram 5.17: 'Flutter Echo' in a modern plastered room

Curing 'Flutter echos' caused by modern plastering

If your room has been completely replastered with hard, modern plaster, the best solution to this sort of replastering is with the careful selection of wallpaper. The idea is to put as thick an absorbent layer onto the plaster as possible.

First use a good lining paper and in severe cases use several layers to build up a good depth. Then choose a good thick expensive paper with a 'flock' pattern. This is rather old fashioned and may not be your first choice but aesthetics must come second to sound quality in the music/Hi-Fi room! An alternative is an 'Anaglypta' paper which is thick and heavy but with an embossed design that leaves a pattern of air bubbles that is very effective in high frequency absorption and the control of reflections. The Anaglypta paper can then be painted with emulsion paint.

Use only matt emulsion paint. I emphasise this point since an enthusiast phoned me complaining that he and his wife had just redecorated their listening room and managed to ruin the sound. All they had changed, he said, was to use vinyl silk emulsion over the original matt. The result was now a very bright, hard sound - what could I suggest to correct it? He knew the answer, of course, he just needed a scapegoat (me!) to take the flack. "Russ Andrews says we've got to redecorate all over again with matt paint this time." Who can blame him?

I tell you this story as an object lesson in how important an apparently small difference can be. The reflective difference between matt and vinyl silk emulsion is easy to see but not easy to equate to the difference in sound it can produce. If you consider though that the small difference is multiplied over the four walls and ceiling of the room you can appreciate how influential it might be. I say 'might be' because it is only a part of a complex relationship involving the room structure, decoration, furnishing and loudspeaker (to name only the principal ones!). Remember that the objective is to achieve a natural balanced sound, rather than pseudo-scientific absolute acoustic values.

Using hangings to cure 'flutter echo'

If redecoration is not an option, Oriental rugs, carpets and tapestries make excellent 'unobtrusive' acoustic absorbers in listening rooms. They are easily hung from rods or specialist supports. If your chosen hanging is really too thin for good acoustic effect you could have a thick backing sewn on. Shops are quite experienced with this and will often ask if the hanging is wanted for its visual or acoustic effect. Where an attached backing is not possible, a sheet of insulation board cut to the right size can easily be screwed to the wall behind the hanging. The insulation board will be invisible but very effective indeed.

Insulation board is a soft, fibrous board used in the building trade. It often has what is called a 'painted' or hard side and a soft fibrous side, allowing some fine tuning of the effect, but normally place the hard side to the wall. Commercial acoustic tiles are made of insulation board and the large panels of shop/office suspended ceilings use a thick version. Insulation board is 1/2" (12.5mm) thick and available in 8'x4' (12.4m x 1.2m) sheets from builders merchants or wood merchants for about £10 a sheet. It can be cut very easily with a Stanley knife.

A very good friend of mine suffered from flutter echo down the whole length of his new enlarged listening room and the cure required treatment of one end wall and the whole length of a side wall! He was understandably reluctant to implement my solution. I recommended that he find a sheer light cream material to wrap round each 8'x4' (12.4m x 1.2m) insulation board. Screwed to the wall, butted together, paintings rehung and they just disappeared! The acoustic effect was a complete cure of the flutter echo problem without any hint of overdamping. The room now sounds very open, spacious and natural. Fortunately it is a very big room and could cope with all the extra absorptive material.

Remember that the balance you need to achieve is of a slightly 'live' overall character whilst the listening end is a little livelier compared to the speaker end. A very useful benefit of buying some sheets of insulation board is to use them initially as temporary moveable absorbers. That way you can quickly get a 'feel' for the effect of treatment in different positions.

Rattle - 'The Sitting Room Chorus'

Way back in the early 1970s, when I was a Hi-Fi dealer in Edinburgh, I had a salutary experience one day when installing a new Hi-Fi system in a customer's house. As usual, I had set up and tested the system in the shop dem room so I knew it was working perfectly - but installed in the customer's house it sounded terrible. Not just bad - faulty! I couldn't believe it.

Had the stylus dropped out? No. Was the cantilever bent? No. Mystified, I put on the HFS 69 test record I always carried and found that the system sounded fine at a very low level but as the volume increased the gross distortion effect appeared. As it was just a single frequency tone signal, I was able to hear some very nasty noises coming from various parts of the room.

On investigation, I found that the nasty noises were coming mostly from a gas fire and a china cabinet. We silenced the gas fire with a thick blanket (it was unlit and cold!) and then went to work on the china cabinet. Not only were the glass panels in the cabinet rattling, but the cups and saucers were rattling too! We fixed all the rattles and the sound from the system was fine. An important lesson learned. I'd discovered the 'sitting room chorus'.

A few tests in other rooms proved that this was not an isolated, one-off problem. It was just more severe than most. Since then I have identified all manner of room rattle and vibration problems that degrade the sound of a Hi-Fi system and have found ways of curing them.

There are basically three categories of rattle culprits:

1. **The structure** - walls, floors, ceilings, windows and doors;

2. **The furnishings** - cupboards, radiators, gas or electric fires, cabinets, boxes, tables and chairs;

3. **The ornaments** - china, glassware, pictures, lights and light fittings.

I hadn't noticed the problem before because the rattling objects vibrate in time with the music and so sound as if they are coming from the loudspeakers not the room itself. This is one of the reasons why a system can sound so different in different rooms. It was also a surprise to hear how loud they were in comparison with the music that caused them. An apparently quiet bass frequency can cause a loud rattle up in the midrange or treble.

So the rattles are polluting the sound of the system even when it is played at low levels. Of course, the more powerful the system and the smaller the room, then the bigger the problem. First you rattle the contents of the room, then you rattle the structure - walls, floor, ceiling, doors.

Finding the Source of your Rattles

Using a signal generator, I found that the stimulus for all these rattles and vibrations was bass energy (frequencies up to about 200Hz). The best test equipment for finding rattles in your room is a good low distortion oscillator in its lowest band of 10Hz to 200Hz, if you are lucky enough to own one, or can borrow one.

Connect its output into the radio or CD input of your preamp, or integrated amp. Then adjust its output and your volume control so that you get a sufficiently loud sound from the loudspeakers that you can make things rattle, but not so loud that you damage the speakers, amp or your ears!

Starting at 10Hz, very, very slowly sweep up the frequency band one or two Hz at a time listening for rattles. They are quite difficult to locate because they seem to come from all directions. You will have to walk around the room to identify the sources of each rattle and probably touch items to stop and start the rattle to be certain what the real cause is. As you locate each culprit, stop it rattling in any way you can (see next page) - if you can't, then remove it!

If you don't have access to a signal generator, there are useful sweep tone tracks on Richard Black's Ultimate Stereo Hearing and Equipment Refresher (USHER) CD - track 34. You should set your CD player on A-B repeat of the first 40 seconds of this track and listen for rattles as the tone sweeps up from 20Hz to 20KHz. Less convenient than being able to have a fixed frequency whilst you identify rattle culprits, but it works.

There is, however, a third way. Nature has provided you with a perfect tapping tool on the end of your wrist - your knuckle is an ideal tapper. You can rest assured that if it rattles when you tap it, it will rattle to the music!

Once you've identified the rattles, there are some very simple, invisible and effective solutions. I will describe the ones I have found to be successful - you may well find more.

Curing Structural Rattles

Walls, floors and ceilings are dealt with earlier in the chapter (see **Room Structure**, page 96).

The most common structural rattle problems come from windows. Sash or casement windows often rattle badly in their frames and can produce very loud noises. You can adjust the positions of the catches to close the window more firmly, but a rubber strip draught excluder is usually the only way to do the job properly.

Small wooden wedges may be the only cure for sash windows because draught excluders are difficult to use on this type.

When you have silenced the frame you will probably find that the glass is slightly loose and rattling because the frame has dried and shrunk away from the putty. The best cure for this is to hook out the old putty and replace it with new - get a glazier in! An architect customer took me to task over a quick DIY tip I gave in a magazine article. He said running Superglue between the glass and the frame on the inside may cure the rattle but it makes reglazing a very much more difficult job. I agree with him so bear it in mind!

If the whole window frame rattles in the wall (I've seen it!) then consult a joiner. My architect friend will probably rap me over the knuckles for suggesting the careful use of expanding structural foam, but that's probably what a joiner would use. You must make your own judgement of your DIY capabilities.

Doors vary in their construction, but most rattle freely in the frame when shut. Draught excluders and catch adjustment will easily cure this. Modern flush doors of light honeycomb construction can often have internal rattles that are impossible to locate and fix. If you have an identical but non rattling door elsewhere in the house then replace the offender with that one, or buy a new one.

An alternative is to line the listening room side with a sheet of insulation board covered in fabric to deaden the rattles. This solution also has the benefit of reducing room 'echo' and the transmission of sound through the door. You may also notice that your room sounds better with the door closed, eliminating echoes from outside the room.

Curing Furniture Rattles

Rattling cupboard doors can be silenced with draught excluders and catch adjustment. Check that any internal shelves are not rattling too - you can stick rubber strip draught excluders on to the supports.

Pine chests or box lids are very prone to rattle but are easily cured with rubber strip draught excluders.

Look out for rattling gate leg tables (when they are folded down) and loose, dried out joints in any kind of table or chair. Dismantling and re-gluing is really a professional job, but I have found a clever DIY kit that allows regluing without first dismantling.

The glass panels in the doors and sides are prone to rattle and it is here that some very ingenious methods must be used to affect a cure. I have a suggestion to offer, but you may find a good alternative. A small piece of Blu-Tack™ squeezed into the angle between the glass and the frame at strategic points will stop the glass rattling and may also be virtually invisible. A rattling cabinet door may be cured in the same way as other doors, with rubber strips and catch adjustment.

Central heating radiators can be silenced with sound deadening pads stuck to the backs. Just slacken off the pipe nuts, lift the radiator off its wall brackets and it will hinge forward to allow you to stick the sound deadening pads on the back. One pad per foot (330mm approx) is about all you need. Watch out for minor water leaks while you do this, wrap some cloths round the pipes at each end to catch the drips.

As I have said earlier, gas and electric fires are a major cause of nasty rattles. They are made up of many metal bits and pieces that rattle against each other with gay abandon. Make sure that they are switched off before you start! Turn off the gas to gas fires and unplug electric ones.

De-rattling fires is made more difficult because, of course, they get hot, so you can't use sound deadening pads, Blu-Tack™ or other dampers that melt or cause a fire hazard. You can make simple and effective pads, spacers and wedges out of aluminium cooking foil and use your ingenuity to find ways of using them to stop the rattles. If you are really lucky, dismantling and then tightening all the screws may be all that is required.

Don't be scared to dismantle these things - most of them are made to be taken apart to replace the 'coal-effect' bulbs and for maintenance (you've heard of annual maintenance of gas fires, of course!).

Curing Rattling Ornaments

Light fittings, table lamps and standard lamps are all rich sources of buzzes and rattles. Routine tricks of tightening the screws up, wedging, packing, holt melt gluing, Blu-Tack™ will all play their part in silencing the light fitting 'sing-a-long'.

Pictures and paintings often have loose glass in the frame, and check that the frame itself is not rattling against the wall. A strip of self adhesive foam or Blu-Tack™ can be used to damp the rattle.

China plates, cups, saucers and ornaments will give you many happy silencing hours! Paper doilies between cups, saucers and plates, etc. work very well (don't forget to put them between the saucers or plates and the shelf!). Ornaments of all types can be fitted with small bits of Blu-Tack™ underneath them to kill the vibrations. My friend Chris found a clever way to stop the Blu-Tack™ sticking to the shelf or table: a dab of talcum powder to stop the 'stickiness' and, hopefully, prevent marking.

Glass benefits from the doily or Blu-Tack™ trick, but items such as bowls and jugs need to be filled with something to stop them ringing. Glass beads or marbles work well. Glasses are best filled with your choice of alcoholic beverage - my favourite is Malt Whisky. Seriously though, stand glasses upside down, the rim is then damped, preventing ringing.

If you have a cabinet displaying china and glassware, do make sure that the glasses aren't rattling against each other. The same applies to the bottles in a drinks cabinet.

Appendix I
Why is RFI such a problem to an amplifier?

A look at the circuitry of an amplifier can be very revealing. We will find a power supply, and circuits that amplify the audio signal. Looking closer there will usually be at least two more circuits. One is called the negative feedback loop, the other is circuitry to protect the amplifier from itself, usually a current limiting protection circuit of some kind.

The negative feedback loop in simple terms compares the output signal with the input signal. If a distortion error or difference is detected, the loop then sends a correction signal that is equal and opposite to the error back to the input of the amplifier. This correction signal is run through the amplifier to cancel the error or distortion. Sounds good, right? Actually it 'sounds' awful. Because this process takes time to happen, the correction signal is always too late to actually cancel the error or distortion. The result is two errors, the original one, and then its mirror image a short time later. This does not show up on normal distortion measurement as the test signal is repetitive. Music is not repetitive, so these time lag corrections show up as added distortion. It should be noted that if the amplifying circuits did not produce the errors or distortion in the first place, the negative feedback loop would not be required.

Now, a look at the protection circuitry. This circuit senses conditions which are dangerous to the amplifier outputs, and momentarily switches them off. This off-on cycling does not in practice protect the output transistors, but causes very audible distortion. Primarily the protection circuit works by sensing DC (direct current), or other abnormal waveforms like RFI or oscillation.

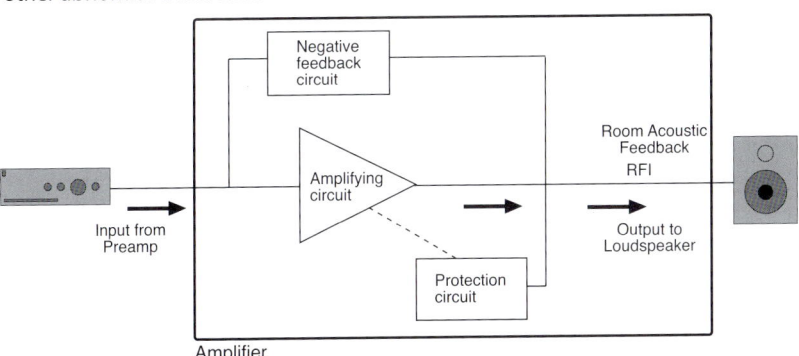

Diagram 6.1: The negative feedback loop within an amplifier

For an overall look refer to Diagram 6.1 on previous page. The arrows show the direction of signal flow within the amplifier. Note that the inputs for the negative feedback loop, and protection circuit are in parallel with the output of the audio amplifying circuit. This means that noise picked up by speaker cables can actually be amplified by the gain of the entire system. As far as the amplifier input is concerned, a signal on the ground plane is just as valid as a signal on the signal input. Noise (RFI) picked up by the speaker cable causes ground plane modulation that will be amplified by the entire system and appear back at the amplifier output with enough power to drive the speaker.

Look again at the Diagram 6.1; notice that there are two inputs at the output of the amplifier. One input feeds the negative feedback circuit that looks at the acoustic Back EMF and RFI noise, and thinks that it was developed by the amplifier, so a correction signal is routed to the input. In other words, noise placed on the amplifier output is recirculated through the amplifier again via the negative feedback loop. The other input feeds the protection circuit.

RFI noise applied to the input of the protection circuitry will cause it to cycle the outputs off because it thinks it's a potential fault condition. Reducing the amount of RFI picked up on the speaker cables translates into preventing the protection circuit from false cycling. This in turn means the amount of usable output power is increased by up to 10dB additional volume without distortion.

Appendix II
The Cost of Leaving Your System On

I have written about the benefits of leaving your Hi-Fi system switched on 24 hours a day, 7 days a week (except during electrical storms) and given a rough estimate of the extremely low cost of doing so. I have bought a mains analyser (for my Silencer and Purifier development work) that allows us to measure the actual current drawn and power used by a wide variety of Hi-Fi equipment. The results are very surprising to those who imagine that their Hi-Fi systems are powerful, current hungry things. Particularly when compared to other household appliances!

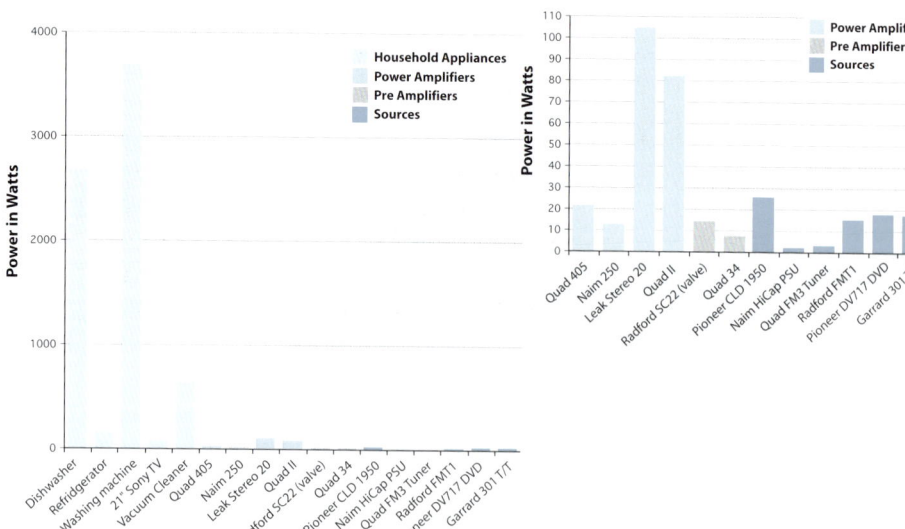

Diagram 6.2 Comparison of the power used by various household appliances and Hi-Fi equipment (highlighting the power use of Hi-Fi equipment).

For example:

- 100W light bulb costs 0.82 p per hr, £1.38 per week.
- Hi-Fi system costs 0.45 p per hr, £0.76 per week
- An all valve Quad system still only costs £2.75 per week.

This represents extremely good value for money. The improvement you will hear in your system performance makes this one of the most cost effective upgrades around!

Offsetting your carbon

Customer Michael Hewitt contacted us to let us know about a website that aims to offset the amount of carbon dioxide we produce (from travelling, heating our home, etc). This is what Michael had to say: "For a number of years I have been following Russ's advice about leaving hi-fi equipment turned on all the time. With all the recent publicity about the impact of energy use on the environment I decided I should try and offset the environmental impact of this practice. I have been able to do so by using the Climate Care website to buy 0.67 Tonnes of CO_2 emissions for £5, which I calculated would be the amount produced by my system (174W power consumption on standby) being left on 24 hours a day for 1 year."

If you want to offset your carbon accurately, you need to work out the power consumption of your equipment – you can do this by checking your equipment's manual, and totalling the 'Power Consumption' figures. Assuming they are in Watts, multiply the figure by 8784 (the number of hours in a year), then divide by 1000 to give a kWH figure for your equipment for one year.

For example, my system consists of a Pioneer CLD1950 CD player, a Quad 34 preamp and Quad 405 power amp. The power consumption in Watts is 55.16W. Multiply this by 8784. Dividing the resulting figure by 1000 gives a figure of 484.52 kWh.

If you aren't able to calculate the power consumption figure, you can make an educated guess. I worked out the power consumption figures for some of my older Hi-Fi equipment a few years ago, and you can use this to make an estimate. As a very rough guide, 200W should cover most Hi-Fi and Home Cinema amps, a preamp and CD/ DVD player and tuner.

Armed with your kWH figure, go to http://www.climate-care.org/index.cfm and click on their 'house' link. Entering my figure for the Quad system in the kWh per year box, it totals 0.36 tonnes of CO_2, and the cost to offset it would be £2.69.

Climate Care then give instructions on how to pay this amount, and the money is used to fund projects that reduce emissions on your behalf, for example, restoring a rainforest in Uganda!

Appendix III
About Kimber Kable

Russ first met Ray Kimber and discovered how effectively his cable design solved the fundamental cabling problems back in 1985. He has been the exclusive UK distributor for KIMBER KABLE ever since, and remains convinced of the superiority of this product range.

KIMBER's cables are characterised by their musicality, transparency and accuracy, producing a sound quality that has won the hearts and minds of reviewers and customers alike.

Leading the field

Based in Utah (USA), KIMBER KABLE have led the performance audio cable industry in design and precision manufacturing technologies for over two decades. Today their laboratory is regarded as one of the best-equipped in the audio industry. The inventive insight of Ray Kimber and his staff combined with their large investment of resources in technological developments and in-depth research ensures that they will continue to lead the field in cable innovation.

The RFI problem

Ray was involved in Pro-Audio sound rigs for concerts when he discovered that RFI was a major cause of bad sound. It was his solution to this problem, and recognition of its application in domestic Hi-Fi systems that brought him into the cable business.

Unique solution

From 1977 to 1979 he developed his original braided cable design and the machinery by which he could manufacture it. In 1979 KIMBER KABLE was incorporated and began producing 4PR speaker cable.

There are now a few other woven cables on the market, but none use KIMBER's patented woven cable design and we haven't heard any we thought could compete with the KIMBER products.

The design and theory behind 4PR was such a radical departure from existing technologies like coaxial, 'litz', twisted pair, parallel pairs, wave guides and so on, that it has taken the past two decades for the technically inclined to come to terms with KIMBER's 'braided' technologies. As RFI is a problem throughout the system, the same basic principles that have made 4PR so successful and enduring are effectively implemented in the products that they manufacture.

Appendix IV
About Russ Andrews

Russ's thirty plus years in the industry, much of it spent in fundamental research and development work, has given him an unmatched appreciation of the complexities of how a Hi-Fi system works as a whole within its environment. The concept that the hardware is separate from but entirely dependant on its 'infrastructure' - the mains, the cabling, the supports, even the room itself - is key to his approach. Basically, he has learnt that you can get really enjoyable music from any level of equipment if you go about it in the right way.

Looking for a solution

An enthusiast from the tender age of 14, he bowed to the inevitable 10 years later by becoming a Hi-Fi dealer (in Edinburgh). He was fortunate to be part of the 'golden age' of Hi-Fi in the 70s, along with his good friends Ivor Tiefenbrun (founder of Linn Products) and Julian Vereker (founder of Naim Audio). However, dissatisfaction with the results given by the accepted Hi-Fi 'wisdom' led him to begin a lone research programme into why Hi-Fi systems don't deliver the promise of recreating live music in the home.

Starting with an investigation of mains quality (pioneering research into this vital area), the research rapidly expended into all aspects of Hi-Fi systems and their environment, and continues to this day. Along the way he worked as a Research & Development consultant with several British manufacturers (including QED, ARC, Nytech and Meridian) in all products from pick-up cartridges to loudspeakers, before setting up his own company in 1982.

20 years at the forefront of innovation

2006 marks the 20th Anniversary of Russ Andrews Accessories Ltd (originally known as Russ Andrews Turntable Accessories Ltd or R.A.T.A.). Russ's pioneering research into the effect that mains quality has on Hi-Fi systems led to his development of the world's first Hi-Fi mains cable (the PowerKord™) back in 1986 and today he is regarded as one of the world's experts in this field.

In recent years, Russ's skills have been in demand in the Pro Audio field including working with the Astoria recording studio (owned by David Gilmour of Pink Floyd), Russell Watson (on his home studio), and respected recording engineers Michael Zimmerling (Simply Red) and Ken Nelson (Gomez, Coldplay).

Twenty years on, his continual R&D ensures his company is still at the forefront of innovations designed to help enthusiasts get the best from their systems. Russ's products have been widely used by music lovers, reviewers and professionals the world over and have won a plethora of international awards for excellence.

Further Reading

Duncan, Ben (March 1997) *Ground Work* <u>in</u> Hi-Fi News magazine, IPC Media

Jackson & Day (1993) *The Collins Complete DIY Manual* Harper Collins

Lawrence, Mike (1997) *The 'Which' Book of Wiring and Lighting* Penguin

Mark, T.E. (1999). *Handbook on the IEE Wiring Regulations BS7671.* William-Ernest Publishing Ltd.

Scaddon, Brian (1998). *Wiring Systems and Fault Finding* Newnes.

Scaddon, Brian (1998). *IEE 16th Edition Wiring Regulations.* Newnes.

Scaddon, Brian (1998). *Electrical Installation Work.* Newnes.

Scaddon, B; Coker, A.J; Turner, W (1997). *Electric Wiring Domestic.* Newnes.

Steward, W E & Watkins J (1982). *Modern Wiring Practice.* Newnes.

Whitfield, J (1995). *Guide to Electrical Safety at Work.* EPA Press.

The Intitution of Electrical Engineers (1998) *IEE on site Guide.* The Intitution of Electrical Engineers.

CEF (1997). *The Inspection & Testing of Electrical Installations.* City Electrical Factors. *(available from CEF distributors Tel: 01229 825952)*

North American readers may wish to refer to the following DIY books that deal with the wiring systems in that area:

Better Homes & Gardens (1997). *Step-by-Step Wiring.* Better Homes and Gardens Books.

Black & Decker (1998). *The Complete Guide to Home Wiring.* Creative Publishing.

Black & Decker (1992). *Advanced Home Wiring* Creative Publishing.

Sunset (2000). *Complete Home Wiring* Sunset Books.
www.sunsetbooks.com.

Useful Contacts

Institution of Engineers and Technology
<u>www.thiet.org</u>

Russ Andrews Accessories
<u>www.russandrews.com</u>
0845 345 1550